The

Hidden Angles of

Life

Professor Sadegh Angha

MTO PUBLICATIONS ®

Sadegh Angha
The Hidden Angles of Life

ISBN 978-190491669-7

A catalogue record for this book is available from the British Library.
Published in Great Britain in 2012 by MTO Publications ®.
Printed in China by R.R.Donnelley

www.mto.org

www.mto-publications.com

Contents

A Note on Translation and Transliteration:

Wherever the masculine gender is used, it is solely for the purpose of linguistic convenience. Since the intent of religion is for the spiritual elevation of each individual, we believe that religion addresses the Soul, and the Soul is not subject to gender classification.

§

Introduction by the International Multidisciplinary Research Association (IMRA)*

In a letter initially written to Hazrat Shah Maghsoud Sadegh Angha in 1973 by Dr. Ronald D. Grisell, sixteen questions of a technical and scientific nature were posed, which have been included in this preface.

This compilation of questions is based on current mysteries which we encounter and which are beyond our control. However, they are of utmost importance in the medical and biological sciences, the advancement of scientific thought, as well as in harmonising the living conditions of beings and the relationship of living organisms with their surroundings; and even in the continuation of human life on this planet.

In the current conditions an immediate need is not identified for resolving these questions. We were certain that as a Supreme, Grand, and Exceptional Sufi (*Aref*) Hazrat Shah Maghsoud Sadegh Angha, could guide those specialists and professionals engaged in researching the sciences. It may seem a little strange to fellow scholars that a

* This Introduction was written in 1975

group of scientists chose this method. Be that as it may, the domain of science is rapidly progressing and developing as a result of direct inner gains and discoveries, which to our eyes sometimes appear as incidents and accidents.

It must be noted that Dr. Ronald D. Grisell has corresponded on behalf of the International Multidisciplinary Research Association (IMRA), whose members all have a specific interest in knowledge and science. That which is noteworthy as an example is their ongoing research and correspondence in various scientific fields.

The aforementioned questions were simultaneously sent to the other members of IMRA, resulting in some very comprehensive and detailed answers. It should be noted that these questions have led to further proposals which will be published in the future. Meanwhile, the comprehensive answer of Hazrat Shah Maghsoud Sadegh Angha, written in an outstandingly original letter of about 260 pages, led to the conception of this current book.

The scope and background of this book

The contents and subjects of this book should be interesting and worthy of attention for readers from any background. This is because Hazrat Shah Maghsoud Sadegh Angha, who has gone beyond the surfaces of the questions, directs us to face their different motives and content. Moreover, in pursuing subjects with higher complexities than that contained within the questions, he elaborates on the following:

1. Sufism (*Irfan*) and its various cultural aspects in Iran.

2. Cognition, which includes the sciences, mathematics and creativity.

3. Sciences and subjects such as: biophysics, neurophysiology, neurology, morphogenesis and biomorphology, senility, death, cancer, biochemistry, physical chemistry, bioenergetics, quantum mechanics, relativistic quantum electrodynamics, relativity, astrophysics, particle physics, nuclear physics, chemical physics, cell physiology, genetics, evolution, social systems analysis, psychology, neuropsychology, psychophysiology and states of consciousness.

4. Meta-sciences and subjects such as: alchemy, homeopathy, the Astral body and its effect on the human being, aura, human magnetics, various kinds of meditation, altered

states of consciousness, healing, telepathy, telekinesis and clairvoyance.

5. Mathematics in the fields of: Artificial Intelligence, Automata Theory, Cybernetics, Creativity Theory and Proof Theory.

6. Philosophy in the fields of: Monism, Existence, Cognition, Perception, Consciousness and Ethics.

About the author

Hazrat Shah Maghsoud Sadegh Angha, also known as Professor Angha, is a Master of Sufism (*Irfan*). He has a school with more than 20,000* students under his instruction in Tehran, Iran. He was born in Tehran on 4 February 1916. For more than thirty years, and also during his time at university, he was under the tutelage of his father, Hazrat Mir Ghotbeddin Mohammad Angha. After this demanding stage, Hazrat Shah Maghsoud Sadegh Angha succeeded his father and took on the responsibility as the Master (*Pir*) of the Oveyssi School, whose roots can be traced back to the Holy Prophet of Islam, Hazrat Mohammad (pbuh).

There are many available books from the Masters of the Oveyssi School; the necessary arrangements have been made to translate some of these into English, with consideration given to their various historical and technical aspects.

In 1974, Hazrat Shah Maghsoud Sadegh Angha was made a member of IMRA in Washington and, through voluminous correspondences with many scientific members of this association, he has become highly renowned throughout the world.

Hazrat Shah Maghsoud Sadegh Angha has extensive experience in physics, particularly in the general theory of relativity, astrophysics

* Based on data from 1975. Currently, M.T.O. Shahmaghsoudi – School of Islamic Sufism® has over 500,000 students in centres located throughout the world.

and astronomy (he operates his own astronomical observatory near Tehran), and nuclear physics, as well as in physical chemistry, biochemistry and Iranian Alchemy (although the latter is not generally regarded as a science in the West). He also has immense knowledge of and experience in the life sciences.

Hazrat Shah Maghsoud Sadegh Angha is writing a book* on the science of Alchemy which will be distinguished from other research works written in the West; since he has based the contents of this Iranian science on firm foundations, using modern compositions and terminology.

In addition to Science and its various fields, Hazrat Shah Maghsoud Sadegh Angha's knowledge and expertise expands to include a large proportion of Philosophy.

Moreover, he has composed several masterpieces of poetry in line with Iran's resplendent history of great poets such as Hafez and Saadi. Without doubt, from different scientific perspectives Hazrat Shah Maghsoud Sadegh Angha is a comprehensive scholar, one of a rare group which has seldom come to be seen in recent centuries.

Comments by IMRA

It is still not clear why publishers, especially when introducing a book, always seek humility or appear to supplicate the reader in

*_Owzan va Mizan (Weights and Balances)_ first published in 1979. It has since been translated into English (Hazrat Shah Maghsoud Sadegh Angha, _Weights & Balances_, MTO Publications, 2005).

apologetic ways. Presumably, the future souls of humankind will be able to do this without any such atonement. However, if we are obliged to do this, then it might be better if we waited and published the work in the future!

In the preface of a recent book, which was initially around 1200 pages, and to serve as an introduction for the main text, the current publisher wrote:

"An important note"

In reading this book, you must be completely sure that you do not go past even a single word without first understanding it. Of course, it is possible that one may delineate the problem that:

Is it possible to align every topic in a completely 'one-dimensional' series, and if such a thing is possible, is this action necessarily the best choice for human perception and understanding?

Therefore, if only one such process exists for someone with regards to solving problems, in particular philosophical and theological issues, and calls one to itself, then the gravity of such 'inanimate' words gives value to reading that book.

As a relatively indirect form of apology with regards to this book, we must say that: if it were not for the thoroughly intellectual and astute actions of individuals, the best advanced notice we could have given for starting the subject would be thus:

> "In reading and studying this book, neither get confused nor be mistaken, nor linger for too long on words you do not completely understand".
>
> This book as a whole must be studied as a series. This is because it is a book which in its own way is creative, and it can be said that in the fields of Creation and Genesis, and their contents, it consists of in-depth predictions on life and how it will be in the next sixty years. Its poetic themes, which have been combined and presented with scientific definitions, have broad connotations, such that:
>
> "For measuring anything, a suitable unit for that subject must be chosen. For example, we say twelve fluid ounces of lemon juice, a gallon of petrol, or two pounds of apples or a heart full of joy and love."

For this reason, the main text and its grammatical translation have been protected as much as possible. Its words, whilst clear and decisive, retain the expression of the meaning and yet have different interpretations. It is for this reason that this school is in the process of compiling a technical dictionary to define the terms in this book, and also the other publications from Maktab Tarighat Oveyssi (M.T.O.).*

At the very least, this book contains two dimensions. Therefore, we must try to avoid using personal perceptions which are derived from (individual) independent longitudinal thoughts (since space is

* In the 1980s, Maktab Tarighat Oveyssi (M.T.O.) formally became registered and also began to operate as M.T.O. Shahmaghsoudi - School of Islamic Sufism®.

probably not flat). 'Levels' is a word used often to describe matters and stages of self cognition and awareness, but nurtures an obscure and narrow one-dimensional path and form of thinking.

Therefore, and only for the purpose of this discussion, it can be said that the logical principles of Existence have somehow found order as horizontal 'levels', in which case, their theories and experiential results are passed on from one to another. (If we say it more correctly: each person has a certain spatial awareness [topology], the dimensions of which change from one point to another). Consequently, we can infer that the degree of spiritual attainment is in a vertical direction. and its successive horizontal 'levels' can be described as thus:

- The State of Science and Arts - this is the starting point which is at the stage of physical life and ordinary leisure (*Qaleb*).
- The State of Illumination - this is an aspect within the spiritual stages of the soul (*Nafs*).
- The State of Knowledge and Cognition - which are at the stage of the Heart (*Qalb*).
- The State of Spirit (*Ruh*) - which is achieved through concentrating on the energies in the heart and passing them through the environment of the heart which ultimately leads to the gamut outside the senses and attachments.
- The State of Secret (*Serr*) -which means cognition: wherefrom the seeker of the truth (*Salek*) will see nothing except God no matter where he turns.
- The State of Hidden (*Khafi*) - where one finds oneself

annihilated in God, and in this case the veils have fallen and the truth of Existence is revealed.

- The State of Concealed (*Akhfa*) - where the *Salek* is not even aware of his annihilation in God.

These are all stages which are considered in Sufism*. With regards to Sufism, really, nothing can be said about it except that it is the final and most complete stage of progress, which has been termed: "Trials and Seekings" (*Seyr va Solouk*).

The words 'equality', 'stability', 'balance' and 'tranquillity', and the phrase 'letting go of attachments and possessions', are most frequently used at the stages relating to this 'Seeking' (*Solouk*). An old saying in relation to this is that 'A Sufi is ashamed in both worlds,' which means no one wants the Sufi but God, or the Sufi has nothing to represent in either worlds.

In the framework of the above conclusions we come across a natural law at the stage of the Heart which either as a reality, leads a person towards the truth (i.e. as a guide), or as an imaginary goal

---

* These stages refer to the concept of *"Lataif"*, originally formulated as "the Six *Lataif"* (*Lataif as-Sitta*) by Hazrat Sheikh Najmeddin Kobra Khivaghi (13th century) and later revised to "the Seven *Lataif"* by Hazrat Sheikh Alaoddowleh Semnani (14th century), (the 15th and 19th Masters of M.T.O. Shahmaghsoudi – School of Islamic Sufism®, respectively). Each *"Lataif"* possesses a specific colour and is associated with different attributes of the Holy Prophets. Inspiration for this "prophetic association" was taken from the famous analogy of Sufism by the great Master Bayazid Bastami (9th century) who stated "its seeds [Sufism] were set at the time of Adam, they sprouted under Noah and flowered under Abraham. Grapes formed at the time of Moses, and they ripened at the time of Jesus. In the time of Mohammad, they were made into pure wine." In "the Seven *Lataif"* the prophet David is also included.

misleads him towards a lie. From another point of view, it is possible to describe truth as the 'Theory of Truth Conformity' when considered at the scientific level and based on its terms and interpretations. (We must bear in mind that the sort of truth derived from deductive reasoning, which expresses relativities in accordance with the axioms, is a different matter).

It is often said that any 'level' is limited by its own laws, which in a way possess a degree of complete or incomplete relativity in relation to similar laws in other 'levels'. However, it is dangerous to generalise this issue. On the one hand, the laws of physics are fundamentally and essentially variable (for example, there is much evidence and documentation suggesting that most of the constant principles of nature and those influenced by gravity are in fact not constant). On the other hand, it may be possible that the more hidden states and stages, i.e. higher than the stage of the Heart, are not subject to laws at all.

In one of his poetic verses, Hazrat Shah Maghsoud Sadegh Angha outlines his vision for the knowledge of humankind in the next sixty years as thus:

Existence itself is in motion

At the stage of the Spirit, knowledge seems to be a connecting aid used to help those at the physical level enter the more extensive stages of self-cognition. However, at the very least, this knowledge accelerates their own progress. Journeying in the heart and passing through it is the last stage, which according to the sayings of Sufis must be followed; and they say that this action must exclusively be taken by 'I'.

Meanwhile, knowledge spreads a language in order to enrich and complement natural language until it relates different cultures more deeply to one another. It is with this language that the ways of eradicating hunger, illness and other sufferings of humankind, which are a great barrier on the course and promotion of spirituality, become clear. It is not unimaginable that in the future the current sciences, which are connected to matter and the world, will become so harmonious with the nature of human existence that inanimate objects will also be used as tools to raise consciousness and awareness, as well as being applied to knowledge of the human mind until more accurate demonstrations and manifestations result from these resources. It is in the more complete stages that 'Logos' (laws symbolising Existence and being) will be discovered and appreciated.

Is this not a form of real prayer and beseeching?

"Therefore, consider every point of this book (or any centrality which the reader has in the universe) from several directions and dimensions, since this method is a path towards studying the sublime dimension." Of course, further dimensions also exist, which in any case totally depends on the individual reader. For example, the fourth dimension can be considered as materialistic subtlety. However, the recurring melodies in the metaphysical systems have a more subtle and delicate existence which influences one another.

A superficial point immediately induces this theory that these kinds of realms (which in other words are entirely materialistic) are completely dependent on the depths of awareness and as such will result in a superficial and baseless reasoning. Based on this, it is said that the stages of self-cognition continue one after another until the depths of hypothetical awareness and observable cognition.

The human Astral body, which is often independently inclined towards the more precise and subtle stage of the Heart and Etheric cover, is in connection with the stage of the Spirit. Therefore, it must be said that with such understanding, the human being immediately seeks to generalise the problems of the inner form. This intertwined issue, which is more precarious than the first, is considered in this complex article.

The details of this problem would not fit in this introduction and we will stay away from them. As such, we shall dismiss them because a part of the solution would inevitably involve the need to match them one by one. However, the degree of accuracy in 'ether' is neither comparable nor equal with the stages of self-cognition or awareness. There seems to exist a 'many to one' ratio between the stages of self-cognition and the degree of accuracy and subtlety. On the other hand, for primary surfaces at least, there is another 'many to one' ratio between the degrees of accuracy and subtlety and a stage of self-cognition. Therefore, in such conditions and circumstances, the independent dimensions are easily observable when they move in different directions and present an organised structure.

To summarise, we must say that: a good law worthy of reference will not confine an individual to the constraints of a limited frame of mind. This issue has been addressed by Einstein in dealing with "observable and referable forms", and has been explained by Hazrat Shah Maghsoud Sadegh Angha at the spiritual states.

The elements of Sufism consist of:

- Remembrance of God (*Zekr*)
- Thinking (*Tafakor*)
- Concentration (*Tamarkoz*)
- Awakening of soul and body (*Sehr*)
- Hunger (*Jui'*) for discovering the truth, inwardly and outwardly, and never ceasing the eagerness to search for more
- Silencing (*Samt*) the mind and tongue, especially with regards to worthless issues
- Fasting (*Sawm*) of the body and mind
- Praying in solitude (*Khalvat*) externally and internally
- Attendance (*Khedmat*), which means: annihilation in the truth of the Master and the reality of Existence, which is God.

From the viewpoint of history, The Hidden Angles of Life can be divided as such:

Before the twentieth century: in summary, knowledge tries to reach the level of and harmonise itself with, religion. The phrase 'to reach the level of and harmonise' here is a complex dilemma in which one endeavours to express others' sayings with one's own words in order to reduce the public disagreement about the known subjects. Therefore, by creating this 'harmonisation', discerning and esoteric individuals, and a generous and honourable Gnostic can pass on new powers to exoteric and superficial observers.

During the twentieth century: as far as we know we have witnessed many confirmations and claims by people who categorised knowledge and religion to high or low levels, or at least at different levels. That is to say it was applied without polemics or debates, and almost in a general and common or public form. During this period, a few endeavours have been made in this field, of which this book is an example. As a consequence of this book, the genuine connections between the inner stages, between the physical and the metaphysical, as well as between knowledge and the Absolute have been identified.

In a confirmation letter which Hazrat Shah Maghsoud Sadegh Angha had initially written upon receiving the questions, he had reminded us that:

"The least benefit which an agreement between Science and Sufism can offer humankind is the discovery of the reality of life through this path. With the aid of God this will happen through our hands, for we are pioneers and innovators of this new scientific method."

The International Multidisciplinary Research Association (IMRA)
August 1975

Foreword

In 1974 the International Multidisciplinary Research Association (IMRA) sent a letter, containing sixteen questions, to Hazrat Shah Maghsoud Sadegh Angha (Professor Angha) in Iran. The answers to the questions, which have been compiled in this text, were translated into English and sent to IMRA.

Preface by Dr. Ronald D. Grisell

Hazrat Shah Maghsoud Sadegh Angha has given me the honour of writing about IMRA in the preface of this book, which comprises the complete answers given to the questions raised by this association. It has been my pleasure to be a member of IMRA right from the very beginning of its formation in 1971. Therefore, it is fair to say that to a certain extent I may be suitably qualified to write this preface.

Just as it has been written in the constitution of this association, one of the fundamental aims of IMRA is: "To establish ways of communication between researchers throughout the world who are aware of and interested in many topics such as life and the physical sciences." Based on this aim, and for the first time, a list of questions about certain phenomena, which are of common interest to all of us and which are growing into a large field through scientific knowledge, were sent to Hazrat Shah Maghsoud Sadegh Angha and later to the other members of the association. (The mentioned questions have been included in the next section of this book for reference).

In addition to personal contact and direct communication, various other means of formal communication which are less common, have been made by the association and it is probably interesting to mention them here.

Currently, the association publishes a biannual journal called Multidisciplinary Research and there are plans for it to be published quarterly in the future. It is to be used as a platform for the unification of different scientific fields and ideas, and also to bring forth these works into the public domain through newspapers. By this means, it has linked different research fields to one another and presented them in a new way. Furthermore, the publication of this journal strives towards reducing the ever increasing barriers in publishing ideas and writings, so as to prevent the many obstacles and expenses involved in accumulating books and journals, which libraries must incur when obtaining and updating their collection.

By paying attention to this problem, we encounter the concept of 'limited edition', i.e. this journal is sent free to major libraries throughout the world for public reference. Through this method most other authors who publish their works in fields similar to the essays of this journal are informed and reprinted copies are sent to them.

As there is no prohibition on receiving a subscription fee for this journal, it is hoped that with more approvals and emphasis on publishing and demand for reprinted copies, some of the expenses borne by its members and non-profit organisations will be reduced. To a certain extent, this issue has been addressed by large scale production and limiting content which is less useful for public reference.

Since, in a way, every journal is limited by its own style and format, IMRA also distributes a newsletter, which was instigated by its members, to provide a means for continuous communication.

The association is also responsible for other specific tasks, such as publishing and distributing results from preliminary research, samples and other classified manuscripts. In the interest of brevity, it suffices to say that the association encourages and aids all novel cultural communications.

In the past couple of years, our small group has grown into a relatively large one, which has made it possible for members to correspond in writing. It is my hope, and that of my colleagues, that IMRA does not lose its character as a result of inappropriate growth and expansion, like many other associations. It is for this reason that every year the addresses of the association's members are printed and published. Our association provides different services to its members, with its ultimate goal and plan being the development of close communications. I also hope to organise regular meetings in the form of conferences, for example.

Due to the association's restrictions and the fact that its members are (interested and attentive) geniuses in many scientific fields - and these people are always rare - up to now many independent committees have been set up at different national levels in countries such as Japan, United Kingdom, Lebanon, Thailand, Germany, Italy and, of course, Iran, as well as others from the Western hemisphere which will not be mentioned here. Therefore, it has not yet been possible to set up a permanent assembly.

Currently, members of the association are presenting their views on the 'hard' sciences such as physics, mathematics and computer science as well as the 'soft' sciences such as biochemistry, biophysics, biology, physiology and neurophysiology. Philosophers, theologians and psychologists also comprise a group. Another group, namely those interested in knowledge beyond general sciences, also fall within the above-mentioned categories.

It is hoped that this short introduction about IMRA does not appear as an indirect request or plea for help, since one of the intended aims was to make a provision for the backgrounds and preliminaries, and consequently to introduce this recent work of Hazrat Shah Maghsoud Sadegh Angha. Another aim was to express our deepest appreciation to him, since this work is an inspirational introduction on the path of communicating and exchanging thoughts and ideas. In reality, there are also other answers being compiled to the raised questions. This book can be used as a guide for future collaborations and publications by the members.

Dr. Ronald D. Grisell

Questions set by IMRA

1 What are the causes of natural death i.e. without considering accidents or severe diseases? Are they possibly the products of spontaneous chemical action-reactions related to positional and collective changes within deoxyribonucleic acid (DNA) molecules, and/or do they occur as a result of the culmination of irreparable damages produced by radiations and diseases?

2 Is death necessary for the gradual evolution of species? This question arises because some living organisms evolve and adapt to the conditions of a changing environment, or harmonise with the conditions of a stable environment, by developing and expanding the internal structure of their cells or by intercellular action-reactions derived from cell reproduction.

3 Why do living beings compete with each other instead of cooperating (or loving one another, in the case of human beings)? This question reminds me of a time when I was standing in a vast field of grass. I asked myself, “Why are these billions of individual blades of

grass living separately from each other?". Then I looked at a jungle and asked, "Why do those trees compete and struggle with each other in order to position themselves in sunlight? Why can they not form a large living colony and cooperate with each other?". It seems that this unity cannot be implemented due to the dispositional limitations of the size of a living being and its need for respiration, nutrition and movement.

In fact, when the size of a living being increases, its volume will increase faster than its surface (due to respiration and nutrition) and even faster than its height (due to the structural strength of bones and thier ability to assist in motion). In an imaginary case, a living being going through evolution could have developed in a vaster area with a relatively smaller volume. For example, if the blades of grass in a meadow could grow in one unit (and therefore a common root would be used by the entire body), it could respire in a more simple way and would not need a strong tree-like wooden stem to hold the branches. Furthermore, if it covered enough area it would not need to move. There is strong evidence for this from a recent study of intercellular communications as a mechanism that reproduces a group of specific cellular structures.[*]*

If it truly happens like this, it seems that this mechanism should also include cancer. What is especially interesting for me is the possibility that this action could be controlled by us and therefore be

---

* B. Strehler, G. Hirsch, "Codon-restriction theory by aging and development", *Theoretical Biology*. **33**, 429-474 (1971).

used effectively in medical treatment, in the use of limbs or organs of the body, as well as the development of the body and the brain.

4 Considering the state and structure of cells, what are the methods of cellular communication? It would be surprising to find that a surgically implanted 'third eye' on top of a salamander foetus' head would develop visionary nerves - with all the neural communications of a normal eye - within the correct hemisphere of the brain of the animal. There must be a factor which guides these neurons in a most accurate and delicate manner.

5 As suggested in Question 4, the cells of the nervous system have great potential for regeneration. By the same token, brain cells have the highest degree of cell death (otherwise known as apoptosis) among a variety of nerve cells, to the extent that in the human brain almost 100,000 nerve cells die per day. How does nature address and balance this matter?

According to the view of one scientist, this matter has lately been found to be connected with learning ability. However, the more important issue, in my view, is that mammals are in greater need of nerve cells than other creatures. It might be that in other animals, the brain is essentially, like other organs, a living organ that balances the internal functions and therefore acts according to the needs and demands of the body; whilst in the case of mammals, the cortex of the brain has greater value since it is the locus for balancing the external activities. It would be interesting to consider, for example, whether, or not, the cells of the cerebellum die at the same rate as the cells of the cortex of the brain.

6 Why do two main branches of living organisms exist, i.e. plants and animals? Is this because only two important issues exist with respect to favourable conditions for their chemical evolution (one being based on cellulose and the other on gelatinous phospholipids - the lipoprotein and cholesterol systems)? Or could it be that there exist 'active and hardworking' parts, or components with great potential and longevity, which are necessarily very complicated?

Plants must convert energies emitted from the sun to food and thus produce ingredients for animals to enable them to move and therefore sustain the plants; and in some cases animals even supply chemical components that plants cannot produce.

To this day our very slow and limited analogue computers do not have the processing power to act like living organisms. However, what would happen if we built computers that operated at the speed of light - or other energies which are a million times greater and more extensive - and which could be used on a larger environment to evolve and actualise many primary 'building-blocks', such as molecules?

7 In reality, are the blades of grass and other living forms on this planet in unity?

Here my intention in using the word 'reality' is as a measurable entity that may be evaluated and on occasions compared with other samples. As for 'unity', I am referring to some forms of living beings or entities who live together cooperatively.

8 Is telepathy a collaborative phenomenon?

This action might be possible by collaboration among brain neurons as follows:

A neuron on its own has no sensitivity towards the electric, magnetic and sound fields, or special fields, emitted from another brain.

As far as my calculations have shown, two neurons which have an effect on one another, and in doing so increase their individual potential sensitivity, will be more sensitive than when they are acting independently. I have calculated through established electro-chemical methods that the sensitivity level of two mutual cells acting jointly is equal to the square of the sensitivity level of an individual neuron. This proportionality is continuous, as collaboration between three mutual cells will have a sensitivity level equal to more than the cube of the three cells acting independently. Similarly, four cells will have a greater effect, and so on.

9 Is there a 'non-chemical' cellular relationship between living organs (i.e. through other fields or telepathy)?

10 If we are truly the epitome of an extraordinary biotic machinery, would this struggle and competitiveness not itself be the avatar of a greater collaboration?

11 Do time and extra sensory perception not have any relation to one another? Please pay attention to the example below.

It is not impossible to believe that in the future we will be able to surgically operate on nerve cells. Imagine that the memories of a pair of twins, Mohammad and Masood, are permanently transferred from Mohammad to his twin brother in a very short period of time (for example around $1x10^{-3}$ sec). As such, the memories of Mohammad will not be effected but the memories of Masood will be wiped out by the transfer. In this situation, Masood's body and mind are in a state identical to that of Mohammad, whilst Mohammad has been transferred to another place, with the intent that Masood will have the same perceptions and feelings as Mohammad would have had before the operation took place.

It is apparent that the original Masood has died during this operation. Now, which one of these active brains is the real Mohammad?

The Mohammad who was in the body of the original Mohammad suddenly finds himself in another place, whilst the Mohammad who is now in the body of Masood does not remember any disconnection with regards to his memories or the place he is currently in.

Is it not correct that the Mohammad who now lives in the body of Masood did not feel any disconnection with regards to his perceptions at the time of this operation?

12 With respect to the general theory of relativity, is it possible to regard perception as a specific intersecting point of space with the fourth dimension of space-time? If this was so, then it would be necessarily dependent upon a specific choice that the observer would have in selecting a harmonising system. Would another observer, observe a different individual? Or, in a way, would the perception of an individual be stable with changes in coordinates?

If perception in this sense is dependent upon the point of vision, then in a way it is possible to change the coordinates, so that a section of an individual would be in union with another section from another individual, in the same manner that the transfer took place in the example of the twins (as mentioned in the previous question). By paying attention to the time of the occurrence of the two incidents, in which one is the receiver and the other the transmitter, we are therefore able to transfer thoughts and feelings from one individual to another. The principal problem would be the determined changes of the coordinates.

13 Should a physical force or field display kinetic activity in the faculty of perception? Or, is this principal related to the inference from changes to the coordinates in the surrounding environment of an individual or other people?

As an example, by adjusting the control panel of an elevator we can reduce its speed to such an extent that it would seem stationary. Or, by modifying the dimensions we can lift a glass of water off the table.

14 From which activities of the brain do an individual's creative powers occur?

In general, this type of action is described as the combination of perceptions from tools and prior experiences in a new situation or action. The meaning of the word 'new' is that we do not use the past experiences of others. Could this action, not be the result of a series of irregular neural perceptions of the brain, which are probably situated in the skin layer of the temporal muscles? Or is this action caused by the high and regulated amplification of some weak and irregular fluctuations?

In the vast field of computer science a programme has not yet been developed which is capable of discovering the logical rules in everything. However, there are programmes capable of carrying out unstructured research using logical calculations. Some number sequences have also been developed which are suitable for proving rules. In any case there is room for hope. Be this as it may, the field of mathematical sciences in general does not seem to show a regular rule and order.

Something which is beneficial in the discovery of one thing may not necessarily be beneficial for the discovery of something else. Thus, it does not seem possible that new rules can be discovered by way of other rules, unless it is done by 'trial and error' or in other words by means of experimentation.

15 It is often said that the most powerful 'discoverer' in mathematics is 'focus', meaning that the proportion of interacting relationships to general conceptions must be within a hypothetical range. Therefore, there is a direct relationship between an increase

in 'focus', or decrease in the number (complexity) of general conceptions, and elucidating axioms with primary conceptions. As such, an increase in the complexity of the interacting relationships between these conceptions, in the shape of theorems or problems, relates them to one another. However, to understand 'complexity' on its own requires a theorem.

For example, imagine that someone (suspended in time) describes the colour 'green-blue' as a new colour; such that until the year 1973 this colour is green and after that it is blue. In the same manner, another person describes 'blue-green' as a new colour, and that colour is blue until the year 1973 and after that it will be green. Now, with the most brilliant description of the complexity of conceptions and perceptions, 'blue-green' and 'green-blue' are more complex in the perception of green and blue, outside of that which they have been described. Whereas, the colours blue and green can be described, as a result, on a higher level of the condition of the 'blue-green' colours and 'green-blue' colours.

Until the year 1973, green is 'green-blue' and then after that it is 'blue-green', and in this format blue will be 'blue-green' until the year 1973 and then after that it will be 'green-blue', therefore we have described a simpler perception in a more complicated situation of perception. So indeed, what is 'complexity'? If we knew, perhaps we could understand how humanity will invent modern mathematics.

16 Is love anything other than a perceptive behaviour? Is this not correct, that there might be a strong force that will make humans capable of communicating via telepathy and curing one another through this manner? Also, many awe-inspiring physical actions may be done through this manner. A colleague who is a teacher of Yoga believes in this principle.

Grand Master Hazrat Shah Maghsoud Sadegh Angha's response to the questions of Dr. Ronald D. Grisell, member of the Governing body and Chief Editor of IMRA

In the Name of God, the most Beneficent, most Merciful

In view of the fact that I had some pending research to complete, I would like to apologise for my delayed response to your original questions. I welcome you to study this rather detailed and extended reply which has arisen from the subtle nature of your original questions.

Since the questions were somehow interrelated, preparing separate answers would have led to frequent repetition and referencing of particular subjects, which is not suitable for a scientific article. As a result, I have endeavoured to give the answers in a comprehensive discourse, which not only clarifies my research technique and experimental methods, but also those of other researchers. However, some ambiguity may arise in translating from Persian to English, which, in this case, will be rectified in any future correspondence.

I hope this article proves beneficial for the discovery of unknowns on the path of research and for progress towards humankind's sublime objectives.

I take this opportunity to wish you and other researchers at the International Multidisciplinary Research Association (IMRA) great success.

Sadegh Ibn Mohammad Ibn Abolfazl Angha.
Tehran, Iran, 4 February 1975.

ONE
INTRODUCTION

Throughout the course of history, at a time when the sciences were collectively known as 'wisdom', it was not unusual for people to study natural sciences, mathematics and theology simultaneously. 'Wisdom' was so broad that any researcher could gather incomplete compendia of information, like general knowledge, on which to focus his thoughts. As this method did not resolve any problems, science was gradually divided into the theoretical and the empirical. This started around 3000 years ago as a result of the ongoing conflicts of the schools of thought during the times of Democritus, Socrates, Plato, Aristotle and the sceptics. This approach lasted until the nineteenth and twentieth centuries. During these two centuries, more than ever, science changed from those incomplete compendia to more detailed descriptions, such that presently every scientific discipline has hundreds of specific sub-disciplines, which are being defined and researched.

From the beginning, knowledge was focused on a luminous point, which should be regarded as the innovation of thoughts in the development and strengthening of the foundation of civilisation.

Every subject, in any case, should have been illuminated by its light. As time went by, two divergent sides, different topics and numerous contrasts emerged from that unique point. Although today the different categories of knowledge are still, to some degree, based on that initial division, they sometimes lose their cohesiveness and interconnectivity due to such extensive expansion.

Naturalists do not pay attention to metaphysics. Their reason for this is that the invisible metaphysical forces, which are a form of the Absolute, cannot be physically measured nor recognised using existing natural laws. Therefore naturalists claim that these forces do not exist or that, at best, they are unaware of them.

Using similar analogical deductions, theologians and metaphysicians from the East and the West have acted in the same manner. Based only on their reasoning, or their own experiments, they believe that because physical matter is limited it cannot be the key to discovering the detailed and infinite intricacies of spiritual cognition. Even if they consider research in the natural sciences to be beneficial for natural life, they do not consider it to be suitable for spiritual cognition.

It is unnecessary to prolong this matter in this introduction, because researchers have realised that the works of the naturalists and theologians lie on two opposing poles and, as such, do not have a common meeting point from which to construct real scientific peace.

Contemporary human beings are dissuaded from looking at the abominable performance of religious rituals. This is, on the one hand, due to the discovery of the cosmic and ultra-cosmic forces which have broken physical laws and limits, and on the other hand due to ambiguous theological principles and discursions in accepting blind faith, which they themselves have termed 'intellectualism'. These, together with the conclusions gathered by human society throughout past centuries led to 'materialism' and 'prohibitive religious ceremonies' as two schools of thought. In other words, from here on, no wise and logical person is willing to follow the irrational principles of the past.

Your attention, as a scientist, to the basis of procedures and your quest to cognise and define the reality of life in a way that, through appropriate measurements, will become established in the field of intellectual research, is the aim of every thinker of our time.

This is because, as you mentioned – and leaving Absolute subjects and their complexity and ambiguity aside – science in the past few decades has to a certain extent suppressed spiritual cognition as a category in the field of research.

In my opinion, the methods used for discovering truths in recent centuries in Europe are also a form of research into materials and surfaces, and modern psychology is an evident example of this.

The path you have set yourself as a scholar - to cognise Reality by attaining complete knowledge of mathematics - is in my view rather difficult but valuable and promising.

If the knowledgeable mind had postulated the already-accomplished cognitions as ultimate, then certainly the last detail of civilisation would have been attained and the truth of nature would have been announced to the world.

However, the true knowledgeable person, whose thoughts are established on disposition, deem an invariable, immutable, and infinite (or annihilated in infinity) centre of character. Starting from this core of concentration, they instantly traverse an infinite and complete realm all around the circle of 360 degrees. Again from each hypothetical point of this realm, they traverse complete curves, thereby displaying the true mesh of perception. While unable to reach the end of this realm ever, at any point, they perceive that tranquillity which is the outcome of those true foundations of disposition. Therefore, they make no mistake in projecting the outside world on the basis of these foundations. The advantage of this spiritual success is that no scholar is childishly astounded by single discoveries and does not consider them extraordinary miracles.

One of the points concealed in your commentary was related to the current state of beings and tracing the history and cause of their existence.

We ought to know that the current existence of beings is a manifestation between two ambiguous worlds; one is the past, which according to philosophers' sayings can be the cause of the existence of beings; and another, the future, which has been assumed to be the conclusion. It goes without saying that the more we discover about these ambiguous worlds of the past and future of a being, the more assured we will be in discovering the truth of life and the reality of Existence.

A point that I must mention is that through our research we should endeavour to reconcile the separation between the material and spiritual realms, which seem to be distant from each other.

It can be seen that the oneness of Existence is repeatedly propounded in the teachings of the prophets. If we separate the teachings of the prophets from the boundaries of ecclesiastical ceremonies, which conceal the truth, and research them in the more subtle spheres of thought, it may greatly help the capable and perceptive spirits to discover Reality.

The structure, movement, behaviour and relativities of the smallest particles - such as atoms, molecules, electrons, protons, photons, and other new-born products resulting from the collision of particles, which, according to the physicists of our time, exhibit dual qualities and appear as wave and particle - to the most immense planets and galaxies have created the structure of the universe. They are the manifestation of the endless, eternal, expanding and contracting wave of nature, which constantly moulds the individuation of particles.

It comes to mind that the wave-particle duality of the elementary particles of nature, which was mentioned previously, is a strong reason for the expansive unity between nature and the metaphysical powers.

It does not mean knowing Existence solely as matter and ignoring the spiritual cognitive forces. Nor does it mean identifying matter, in nature, as superfluous and without foundation. Rather, it means knowing matter as the appearance of Existence, and those sublime forces which in their ultimate state are unattainable through research into life - and in fact lose their shape and dimensions in the Infinite - as being the truth of Existence. At times, these forces, due to a series of interrelated and relative causes, present as natural and sensory manifestations, whilst they have never been disconnected from the Infinite.

I have realised that the existence of every being lies in the chronicles of its events from ‘non-direction’ to ‘non-direction’. Therefore, to become acquainted with each being and to have approximate information about its past and future requires deeply accurate research into its current existence.

In short, any being shows, or at least through research can show, some signs of the factors and causes of its existence. However, in this research, what presents us with difficulties is that we cannot reach all the hidden factors and physical means within a being.

Just as you, I do not base my thoughts on non-intellectual and unreasonable arguments and explanations. When conducting biological research I always try to use the appropriate and correct gauges. Thus, I have come to understand that in order to solve different problems, both physical and metaphysical methods must be employed.

It may not be easy to cognise the source of the commands and induced voluntary or instinctive actions and reactions of beings with similar organs, particularly in the case of the human being who possesses a complex system with intricate characteristics. Therefore, in the research of the scholars, it is not possible to look for definitive laws to use as a criterion through compendia or extensive research.

We will discover the reality of objects only when, while studying them, no angle remains hidden from our research. As a result, if there are unclear and unknown points in the observation of objects then whatever we have presumed to know is definitely nothing more than that which our senses and understanding have introduced to us.

Throughout past centuries, all theories regarding various biological aspects were limited to what scientists could understand and present using their own instruments and capabilities. Afterwards, as the research apparatus gradually improved and personal presumptions grew, definitions and laws also changed.

For example, there was a time when human beings believed the Earth was the centre of the universe, and later it became known that the sun is the governing centre of our solar system. Nevertheless, with all these explanations, from an exact philosophical point of view, the human soul in the infinite universe is, in a way, the truth of Existence and its centre.

It has to be considered how far our senses and sensory perceptions, with their specific and limited range, can guide us towards cognising life. If the universe were finite, then measurements would have solved our problems. However, if the universe is infinite, then limited science will not be able to reach its extremity and a limited unit of measurement will not be able to cognise the unlimited.

According to Kant, beauty, or anything which appears beautiful to us, will not open a world outside of us; rather, it reveals a special state of our spirit.* It is not the properties or characteristics of objects

---

* In his _Critique of Judgment_ (Book I: Analytic of the Beautiful) Kant states: "Taste is the ability to judge an object, or a way of presenting it, by means of a liking or disliking devoid of all interest. The object of such a liking is called beautiful". He explains that the feeling of pleasure and displeasure ("liking" and "disliking") are associated with aesthetic "Judgement" (one of the cognitive powers in Kant's trichotomy of the higher cognitive faculties; "Reason" [or Intellect] and "Understanding" being the other two). Furthermore, in an aesthetic sense of the word, Kant describes "spirit" as the "animating principle of the mind" and the ability to present an intuition which prompts much thought. Therefore, this "special state of our spirit" referred to by Hazrat Shah Maghsoud Sadegh Angha is what Kant describes as "the harmony between the form of the object...with the cognitive powers – i.e. the harmony required in general to unite an intuition [spirit] with concepts so as to produce a cognition." (Immanuel Kant, Werner S. Pluhar; Mary J. _Critique of Judgment_. Hackett Publishing Company, Inc. 1987).

which make them beautiful in our eyes (as these characteristics will remain dark and hidden to us); rather, it is a power in them that can excite our spiritual strengths, especially our intellect and moral virtue, and acquaint, attune and harmonise them with each other.

We do not sense the special melody of objects; instead the tune we perceive is that of our own spiritual strength.

In the book *Hamaseh Hayat (The Epic of Life)*, I have tried to define, in brief, the physical faculties of the human being and the range of its sensory capabilities, according to the laws of physics, and then to draw the researcher's attention towards considering the superior aspects of forces.

We should either entirely ignore the prophets' teachings (I do not mean people's commentaries on and annotations of their teachings) or accept their principles as examples and guidelines for the progress of natural sciences. In my opinion, not to consider those sublime, infinite principles, which have more to do with the aspects of intuitive cognition in the realm of the Absolute, will confine us to the limitations of our senses, perceptions and analogies in such a way that we will be unable to become aware of the truths, even though they are cognisable.

Obviously, you will not be surprised to find out why, in explaining my points of view, I redirect your attention towards the real teachings of the prophets. This is because the main reason behind my research at the start of my spiritual journey was the cognition of religion and gaining knowledge about the realities of Existence. Therefore, I had no choice but to begin my journey by studying existing philosophical concepts. However, what I understood through studying the thoughts and ideas of natural philosophers, mathematical philosophers and theologians led me to divide knowledge into two categories: one, investigating surfaces, and the other, cognising realities. Moreover, during my research I tried not to allow my personal and cultural thoughts to interfere with the cognition of those categories.

Then, I contemplated the significance and necessity of research into the physical and metaphysical sciences. I wondered about a person who continued his physical life and disregarded this type of research: would he then be a wrongdoer, and therefore suffer irrecoverable loss? Or, is knowledge of manifestations a form of groundless egotism brought about by our neurological and psychological complexes?

Thus, in analysing issues pertaining to physical life and real life, I realised that the endeavours of humankind can be summarised into these two aims: man is in search of his absolute tranquillity, and also, because his threatened by death, he should know and apply survival methods. I concluded, therefore, that the main quest of humankind, in diverse categories of life, is firstly, tranquillity, and secondly, survival, that is to say to live in peace and to be eternal.

In the School of Islamic Sufism, which is based on the real

cognition of life and Existence, it is declared that life is omnipresent in its perfection and strength. Thus, it is not wrong to say that life also exists in death. In order to understand death, is it therefore not preferable for us to pay attention to birth? Since we have all experienced birth we are able to understand it better than death, which we do not know much about so far.

It has been said that the first particle of life originated from the seas some two billion years ago. Was this living phenomenon really the first manifestation of life? Or, could it be that, prior to this, life had already existed in some form and human beings only became aware of it at this stage of its manifestation?

Biology has carefully tried to follow the life journey of such a 'particle' in order to discover life and approximately know its history from the time of its initial mysterious form to its subsequent births. This type of research, which started in 1606 in the West, has continued until now.

Dr. Loren Eiseley, of the University of Pennsylvania, asserts in his research that we should look for the wonders of our world in water because life originated from water.

From a scientific point of view, one of the aspects of 'living' can be described as: a physical state in which energy and matter are constantly interacting and interchanging. This state is defined by manifestations possessing extraordinarily mysterious and complex structures. A living organism is constructed from compositions of complicated macromolecules which communicate with each other in a well regulated manner. Such an organism must obtain energy from the outside environment and prepare the required mechanisms for its life using this energy. Moreover, 'living' necessitates perpetual changes and reproduction for its continuation.

How effective can cell structure - consisting of the membrane, internal and nuclear molecules, deoxyribonucleic acid (DNA) and ribonucleic acid (RNA) compositions and so on, and considering all of its composing factors - be in discovering the visible and invisible forces?

Consider how blood from the pulmonary capillaries around the alveoli is involved in chemical and physical actions and reactions. Is it not true that nitrogen and oxygen, which constitute the air, refine the blood and take rich reserves of solar energy from the air? In fact, through this act of refinement the forces of life are returned to the heart, wherefrom they are sent to all other organs. Is this not an organisation of cosmic life? How should we analyse this action-reaction process in order to cognise nearly all the secrets of life?

On the other hand, in discovering the origins of cellular life, it can be observed that living organisms started their lives from the actions and reactions of hydrogen-oxygen based compositions. From the relationship between the above two subjects, should you not pay attention to this fundamental and authentic law of life?

According to the ancient sages, mercury was the 'mother' of all metals and sulphur was the 'father'. Regarding their properties, sulphur is warm and dry and mercury is moist and cold. They were searching for a catalytic agent, mainly salts, to act between the two and produce new metals.

This principle is still acceptable in alchemy today, that is to say, the substance must first be calcined and then tinctured so that it can be transformed to a higher quality. It is said that whenever these variable elements are able to withstand high temperatures, and in fact remain constant in a higher state, then that metal would have found its perfection.

At this point, one is reminded of the ideas of the learned who believed that life and spirit were air, respiration and blood. However, today it is well known that air and blood are two intermediaries of life, among various intermediaries, by which, with other similar factors, the Source of Life is able to function with the world of life, via a close and all-encompassing connection. It is from its origin that the Source of Life excites life in a disciplined and regulated manner, and governs the physical body.

If the materials forming the cells, such as hydrogen and amino acids and so on, lack energy reserves, and lack the attraction and repulsion of individual and cosmic radiations, are they dead or alive? If they are dead, then how do we define life; and if they are alive, is life the cell itself or not?

Even though effects are due to effectors, they are not similar to each other, so that one can be known by the reasons of the other. All the movements of the universe are an effect of the truth of life. Therefore, even if all the effects of life could be defined, perhaps the complete cognition of the truth of life would not be without any rational errors.

Humanity is in search of and endeavouring to discover the laws that will solve the problems of life, and enlighten its ambiguous points in such a way so as not to face any scientific dead-ends in its research.

It is nothing new for human beings to be thinking about death and wanting to discover its secrets in order to strengthen and safeguard their own lives. I do not know whether it is necessary that I describe the history of genesis - or that of cellular life, which has been pursued through the self-sacrifice and perseverance of hundreds of researching scientists - and also explain the glorious epic of science, which has travelled a very long way, but is sitting perplexed, far from its research destination, not knowing which of the many ways to take to reach its goal and objective.

If we rely on and respect the valuable research of scientists, we will encounter some groups that will ask us what we know about the truth of life. However, if we do not want to pay attention to scientific experiences, our illogical suppositions will not untie the knots of the problems of life either. Nevertheless, we cannot hopelessly sit with our hands on our knees and surrender life to an ambiguous fate, because this will not satisfy a human being. Therefore, a general study of the initial and present structure of the cells and their components is necessary.

TWO
THE CHEMICAL COMPOUNDS OF PRIMORDIAL ORGANISMS AND VARIOUS SCIENTIFIC VIEWS ON GENESIS AND CREATION

Naturalists and biologists believe that the materials which form living organisms originate from the chemical compounds of the Earth, and in order to gain knowledge about living matter it is necessary to know these materials. This was the basis for the life journey of the structure of particles, from atoms to living matter. Laboratory experiments in this field, although at an early stage, imply a definite involvement of physics and chemistry in the manifestations of life, because in this way scientists have been able to assemble the primordial forms of life, using the combination of simple materials and suitable living conditions.

Atoms are combined together through their binding energy to form molecules, and molecules combine to create macromolecules, some of which may contain hundreds of thousands of atoms.

The molecules of water, which are composed of hydrogen and

oxygen, constitute 75 percent of living beings. In addition to lipids, which are among the essential factors for magnetic currents, this percentage indicates the importance of water in the living structure of the cell. Water conducts life's electromagnetism. One should not ignore the fact that at least 75 percent of the human body is made of water. This in itself is a powerful and efficient magnetic field, which compels the organs within the human body, and all their cells, to perform their functions.

In addition to the points mentioned above, there are five fundamental groups within the principal structure of beings, which are considered to be very important in linking, converting and transforming molecules. These five groups are: glycerine, carbohydrates, amino acids, fatty acids and nitrogenous bases. Combinations of these molecules produce macromolecules in the forms of:

1. Adenosine phosphates – they are the energy reservoirs for all beings.

2. Lipids – they are produced with the help of adenosine phosphates and, like carbohydrates, supply the energy for the body of a being.

3. Proteins – these are the primary structural materials of the living body and also have a significant role acting as enzymes in accelerating chemical reactions.

4. Nucleic acids – which are made up of nucleotides are responsible for all the 'hidden' and transferable properties from parent cells to daughter cells.

5. Polysaccharides – these are composed of glucose and other carbohydrates and play an important role in the structure of the body of a being.

The issue of replication and division, and amino acid synthesis in such a way that 'offspring' amino acids are identical to 'parent' amino acids, also involves important points. The most important one is the transfer of individual characteristics from one entity to another entity, about which many raised or un-raised questions remain unanswered. Following this, we immediately encounter the theory of 'mutation' which accounts for millions of new varieties.

The important issue in amino acids is the stability of their chemical properties, which resist harsh and unfavourable physical and chemical reactions.

Biologists believe that beings have come into existence from accidents and incidents at seashores, due to the appropriate positioning and constant interaction of molecules. Based on this theory, they have even been able to reach some preliminary results in laboratories.

With regard to nutrition, it is believed that initially nutrients were simple materials, and solar radiations transformed minerals into organic compounds.

Respiration utilises the absorbed vital materials by supplying the body with oxygen and disposing of carbon dioxide. Actual respiration, which occurs in the cells of the body, produces and stores energy.

The phase of cell growth and development, which is conducted by the direct involvement of the vital molecule adenosine triphosphate (ATP), commences with the absorption of nutrients. Through this growth and development, the cell is able to repair and compensate for any deficiency within itself. Therefore, all these functions are under the direct supervision of the amino acids.

THREE
CELLS

The first seemingly living phenomena were molecules of amino acids which came into existence from within chromosomes. Upon leaving their membranes, they were scattered into the sea, and, due to the forces of attraction, the interaction between these different particles gave rise to new amino acids. The creation of viruses is assumed to be based on this theory.

The release of nucleic acids, as a result of intercellular transfer, leads to the loss of some of their characteristics as they enter another cell. Therefore, in addition to the two factors of gender and mutations, the following methods have been selected as a necessity for the continuation of life:

1. Heterotrophism – where nutrition is obtained from the body of other animals and takes place in three forms: Parasitism, Saprophytism and Holotrophism.
2. Autotrophism – where beings obtain simple and organic compounds from minerals in the following two forms:

a) Chemosynthesis – where useable organic compounds are formed from the mineral salts entering the cell.

b) Photosynthesis – where nutrition is carried out by simple elements absorbing solar energy and subsequently converting them into organic compounds.

What is a cell?

How is it born?

How and why does it live and die?

Due to the limitations of scientific measuring apparatuses, the laboratory experiments, experiences and results of scientists endeavouring to study the cellular life cycle have only been able to help, to a certain extent, to identify the products of the life of the cell. For example, various experiments concerning the cell cycle show that cells do not always divide into two equal parts, and also, in some cases reach full growth either gradually over a fixed period, or suddenly.

In this journey, the cell, prior to its preparation for the phase of division and in accordance with the law of Existence, endeavours to take the shortest time to reach a point similar to that of large cells (and this is not dissimilar to the development of a foetus).

As you are aware, this cycle continues until the cell has fully grown and is born. Therefore, if a cell fails to divide, it is inevitable that there is a barrier limiting the growth of its size.

It should be noted that natural scientists in the past proclaimed that a specific ratio exists between the mass of the cytoplasm and the nucleus of the cell. However, this theory was later rejected by other scientists. Nonetheless, even if this theory does not seem to be logical, the fact remains that the mass of a cell is limited and controlled by the capacity of any nucleus which supports its growth.

The issue which has to be researched is: what factor sustains and controls the growth of the volume (i.e. the intricate surfaces) of the nucleus?

The duration of cell growth differs for different cells. Under normal environmental conditions, such as balanced nutrition, suitable temperature and respiration, which is itself a part of nutrition, similar cells have similar cell cycle durations. It is possible to prolong this duration under unsuitable conditions, but it is not known how to shorten this duration in cases where one wants to speed up cell growth. However, in my opinion, applying some energy to the vital structure of a cell, in accordance to the cell's aptitude for acceptance will accelerate its growth.

With regards to respiration being a part of nutrition, the proof of this matter is the situation of the foetus in its mother's womb, where feeding and respiration proceed simultaneously through nutrition.

By paying attention to the numerous experiences of scientists it becomes apparent that a cell does not need to divide in order to survive. Indeed, it is the will of the cell itself which is the criterion for division. This is due to the fact that at the G_1 phase of the cell cycle the cell uses a 'hidden key' to enable entry from the G_1 to the S phase. Therefore, these are the types of reported scientific experiences considered by researchers.

An important issue during interphase is chromosome replication. During this complicated and mysterious process each of the 46 strands of chromosomes in human cells makes identical copies of themselves according to a specified programme. After this precise process the 46 thin strands of chromosomes, with a total length of six feet or more, are packed into a nucleus with a diameter of four ten-thousandths of an inch.

This process represents a well-organised and calculated co-operation, since one chromosome will not start dividing until another has completed its task. It is for this reason that the subject of 'completion and continuity' goes beyond an individual activity and affects chromosomes as a group. The function of ribosomal RNA (rRNA) still remains one of the cell's secrets. As yet, none of the theories have been able to fully justify the reasons for the presence of RNA and proteins in ribosomes, or to give reasons as to why a ribosome attaches to a particular site of RNA from where the protein chain starts.*

Another subject of great importance is related to the three groups of enzymes that function in the human body:

1. Food enzymes – come from our diet and help the digestive process. They follow a constant proportional law and increase in a predicted manner similar to that of cell growth.

2. Digestive enzymes – which increase abruptly at certain stages of the cell cycle or will grow in various stages of the cycle.

3. Metabolic enzymes – which increase at certain stages and then decrease, as though they have been programmed for a specific task. For example, enzymes which only appear during the S phase of the cell cycle to synthesise DNA structure.

The existence of enzymes and their functional processes corresponding to genes raises these questions:

Is a cell's production of any enzyme the end result of a pre-planned

---

* In the 40 years since The Hidden Angles of Life was written (written in 1972, first published in 1974) scientists are still unsure as to the role of ribosomal proteins in the formation and functioning of ribosomes. AV Korobeinikova, MB Garber, GM Gongadze, "Ribosomal proteins: structure, function, and evolution" *Biochemistry (Mosc)*, **6**, 562-574 (2012).

genetic programme? Moreover, do genes determine the phases of the cell cycle?

With regards to hybrids in different cells, are there factors which stimulate the progression of a cell from one phase to the next?

To answer this question, phase G_1 and S for these hybrids have been considered and it has been observed that the nuclei of the cells at phase G_1 are starting to prepare to build DNA and in phase S they start to build DNA. Therefore, it has been concluded that the cell at phase G_1 contains an unknown material that delays the production of DNA and that this material does not exist at phase S where DNA synthesis begins.

These experiments and others performed by biologists prove that there exists a governing force which pervades the cell and is responsible for initiating chromosome propogation. Therefore, knowledge of this unknown could be an important discovery for the subject of cellular life.

Moreover, for example, there is a factor in the M-phase of the cell cycle that forces the nucleus to contract, despite what it desires, and if this factor is composed of special kinds of molecules, it should make scientists consider an important issue, and that is: examining these unknown factors in cellular life can shed light on other complex cellular issues and remove ambiguity from research. For example, it is asked:

"How, and with what communication mechanism, are the events occurring inside a cell influenced by the external environment?" and "Could this be related to a specific understanding of the cell membrane?".

It becomes apparent that these cases or others such as nutrition, division and cell birth appertain to a time before the existence of the cell in its current form. This issue relates to a qualitative aspect and a concealed communicating signal within cells. In my opinion the organised interaction during chromosome division has a form of 'natural telepathy' within itself.

In all these explanations the active and intellectual powers are the agents of true births, which in a cell are known as the products of cellular life.

Cells and components of living objects are condensed energies. If for no other reason, this truth should be accepted on the basis that the life of every particle depends on its aptitude to accept energy and to continuously assimilate and repel. It could be said that it is these very energies that in various phases, like waves with different frequencies, have performed their set tasks and, with their balanced interrelationships, have drawn the picture of life onto the canvas of the infinite universe.

That which eats, breathes and shows chemical and physical signs, is born and dies. Each and every part and whole of the world eats and is eaten. They eat in order to be born and are eaten in order to die.

Death is the prelude and the cause of every birth, as birth is the prelude and the cause of every death.

If it were not for the forces of attraction and repulsion, the effects of life would not have existed, at the very least not with this shape and meaning. What would happen if things did not appear as they are in our realisation?

Consumers and those which are consumed, neither take anything away from Existence nor add anything to it.

Death occurs gradually and without ceasing. It is caused by a disruption to the balance of cellular life, which is brought about by a disharmony of various bodily organs with one another and an imbalance in the attraction and repulsion within cells. These arise, due to the unfavourable influence of visible and invisible forces, whether directly or indirectly and without pause. The death which occurs at a specific time for individuals is a more apparent indication, the result of which is a rapid change of the status in the life journey of cells.

These disorders are possibly triggered by one or several localised irregularities which, if not treated in time, will result in an altered state of the cells or bodily organs - this means death.

The word 'disorder' which we have used here is a general term. Its reality must be sought in the life journey of the cell, and the totality of its body and mechanisms, as well as the attraction and repulsion of the internal, external, cosmic, and ultimately, life forces.

The dimming or absence of light from a lamp is caused by the deficiency of electricity or a power cut. This example is applicable to the first question: "What is the cause of natural death, without considering any accidents at the time of death?".

Industries, technical advances and discoveries are always in agreement with natural principles and laws. This means that if a system is not in concordance with the laws of Existence it is incomplete, or in principle, void of any scientific, logical or rational origin.

Of course, one should not assume that an instantaneous answer to this question was revealed to me; indeed, no. Instead, through experiences, I have noticed that in old age and through general fatigue human organs lose their co-ordination and harmony, and even their power, balance and capability, until the minute when the settlement and overall connection between all parts of the body is collectively cut off from 'life's electricity', causing a great change in one's status.

The cause of cellular aggregation in the entire human body makes us notice this active practical and intellectual power. If human life was born out of an aimless accident and its birth, death and life-span did not attract our attention towards an aim, then scientific work on cognising the cellular structures would have come to an end. However, as you can observe, even with all this valuable research being conducted, it seems that the human being has not yet taken the first step in discovering the truth of life and knows nothing about life except its signs and effects.

Death, as considered by religions, is the separation of the material body from the Absolute identity of the human being. This, does not have an observable definition within the confinements of the body's natural deaths and does not always coincide with natural death.

Death, or the second birth, is the complete purification of the spirit in the spiritual realms, and the quest to separate the real identity from all carnal tendencies, which are known to be voluntary or involuntary and which will be discussed in another treatise.

What does cell propagation mean with regards to increasing an Existence which is vast, eternal and timeless? In the realm of Existence, nothing can be added or taken away from what is already there. In order to solve the puzzle of life, first we need to find the functional similarity of the forces acting in hydrogen in our bodies and those responsible for the sun's radiation.

The sun's forces are stored behind the magnifying power of possibilities and the existence of beings. The human entity at equilibrium manages and regulates these energy fields in the domain of capabilities of matter, cells and organs of the body. In this case, it should be noted that the components of the cell are something along the lines of inanimate objects, and the spirit of life is a collection of energies which is active and the cause of every action and reaction in life.

Of course, with this explanation, we are not saying that energies are anything other than the truth of matter. Moreover, at the same time, we do not believe that energies are the ultimate goal of natural life.

Even when, at the very least, the active energy fields of cells and natural particles are clarified, the issue of an intellectual source and driving agent will still remain ambiguous and unknown. For this reason, one should not imagine that cognising the effects of cells, particles and the energy in our universe will be worthless. Rather, in my opinion, to make it possible to better know and clarify the relations between an individual and the whole universe we should discover the characteristics of each level one after another and put those behind us. Consequently, the ceaseless endeavours of humanity throughout the ages would reach a final destination of true peace and eternal life.

FOUR
MAN AND BEINGS

This discussion is continued in the third question:

"Why is it that some groups of species (organisms) are in competition instead of cooperation?".

Now, if we address the question in another way, such as: "Why do parts of beings accept each other's effects, and continue life by following the law of 'survival of the fittest'?", then it immediately comes to mind that all parts and objects have chosen methods of cooperation for their life. A particle loses nothing when it is consumed by another. When billions of plants are living apart from each other, they are still living. You can observe that they are all agents of life. Any particle which falls under unfavourable conditions will dissolve its previous life into another form of life, from which point it continues its life.

In my view, it is better to ask why cellular life lacks competition. In this case the reason is that in the infinite world, because of its unity, contrasts and antagonisms cannot exist; and this is what has been considered in the third question.

Volume is a collection of surfaces. This is because natural or metaphysical energies must penetrate all geometrical points of surfaces in an equal and balanced manner to such an extent that surfaces, due to an absence of quantity, contain all the qualities of energies. As a matter of fact, this in its own right is also another manifestation of natural life. Therefore, we have to be so certain of the vastness of surfaces in Existence that its presentation becomes a criterion for the total Infinite Existence.

I think the above theory would be accepted if we considered volumes carefully; how else could atomic and molecular components become visible if surfaces were not identified? Since the indivisible particle has been rejected, this necessitates the acceptance that volumes are the very geometrical points which inevitably form surfaces. Therefore, the dual manifestation of particles, known as wave-particle, should be considered as a logical, scientific and correct law. For this and other reasons, the membrane of a cell is able to attract and repel energy. In my opinion, the cell contents are full of such geometrical surfaces.

Intellectual energy is greater than that which we envisage as having an exclusive and limited centre in any object, large or small. This is because whatever comes to our senses is but a small portion of the knowledge of Existence. Radiations, energies and their effects, the attraction and repulsion of particles in carrier waves, the points and surfaces which can be sensed - these are all a small fraction of the grand organisation of creation and its manifestations. The organisation of the infinite universe, the small particles and their components, and waves are all in the same way manifestations of life. Furthermore, in respect of the actions of the infinite power on one another, they do not possess any more or less of this knowledge, since equality and balance are the laws of the vast Existence. This is the rule for the existence of life in all the manifestations of life.

With these explanations, sometimes one thinks that perhaps our world has also been preserved in the Absolute figure and form. Yet, our contractual agreements for measuring have concealed their infinite realms from our senses and understandings to such an extent that, for example, the creating agents of water are hidden and liquid will present itself in opposing and contrasting forms such as steam and ice. Therefore, all other elements should be considered in the same manner; that is, although they may have different manifestations, indeed, they have never lost their genuine essence and nothing is added or taken away from them. Objective manifestations, according to philosophical principles, do not have a genuine existence.

Which cell or molecule would nourish or endeavour more than its living aptitude? None, as they would lose their balance of life.

The unity of Existence does not allow us to consider the existence of beings as somehow deeply isolated and separate from each other.

The manifestation observed in a sapling, in which branches and separate leaves - although on the same branch - pursue their path of life, is this unity in Existence that leads beings on their course of life. Existence, which is pure and vast, in accordance with such concepts, is the only nurturing agent. The manifestations of beings are analogous to the branch and leaves, so that while they seem to be living apart from each other, they are in fact connected and rely upon an intellectual and unique agent.

The world of creation is like a canvas, which would have no splendour if its painting was mono-coloured and monotonous. It is the variety of colours and shades which distinguishes human beings, animals, plants, inanimate objects and their characteristics from each other and makes them praiseworthy. If, for example, the painted portrait of a man was created using only one colour, or if all his organs were similar to the eye, which is the most sensitive and exquisite organ, we could not apply the word painting or portrait to it. Conversely, each organ is presented with an attractive tenderness and in a proportionate shape – a result of the variety of colours – and thereby portrays the mastery and power of the Artist.

The beneficent powers and needy beings are nourished by a common source. This balanced association is possible due to their unity and uniformity, which is a result of innate capability. Otherwise in terms of individuality, objects, species and gender have no direct relation with each other.

Through my studies of different topics, I have concluded that the words 'imperfection' and 'perfection', used in natural, mathematical and theological philosophies are two illogical and unreal terms. How can imperfection or perfection be conceivable for a constantly changing manifestation?

If we wanted to add a small particle to the universe, where would we find a void, and if we decided to deduct something, what would happen to order? If these two things happened, the world of creation would be destroyed and become non-existent. Furthermore, what would be the true gauge for measuring imperfection and/or perfection? We neither have a word nor a reference for an opposing realm to Existence, such as 'non-existence'*, wherefrom we could find beings from 'total vacuum' and unknown places, and give them form in the realm of Existence.

It seems that the existential basis for beings is just as you have proposed. However, these manifestations have an unfamiliar and separate semblance when we take the senses and existential necessity into account.

---

* The phrase "non-existence" is used to describe the opposite of Existence until Existence can be understood, otherwise Existence has no beginning, no ending and no limitations. The word 'nihility' has not been used since 'nihility' itself is a part of Existence bearing no manifestation.

I think life and its necessary continuation is impossible in the form of individuality.

Now, if we suppose that individual competition between species is genuine and real, i.e. there is definite competition, then how can we believe in and be certain of universal order within an Infinite Existence, or even in the limited cosmic elements? Well it is obvious that because of the differing needs, conflicts between different factors in the existing environment would have resulted in the extinction of Existence. Whereas we see that this is not the case and, therefore, inevitably the principle of cooperation and balanced capability is more accurate, since beings in the realms of Existence obey an infinite governing law for their life continuation and living manifestations.

We cannot give a verdict that life only exists in two groups of animals and plants, as this would suggest that we have a particular definition for life that is only applicable to the states of these two groups.

What about the life of those plants which decay in mines thus forming coal and producing diamonds from the existing carbon? We should acknowledge that the acceptance of effects is one of the definitions of life and so we have to extend the concept of life to the realm of beings in different groups.

Once we are certain that the cell membrane is composed of

proteins and lipids, and that the inner composition of the nucleus and the area in between the two is also of the same known materials; and that the cell membrane can expand and proliferate its size several times by nourishing and nurturing, then we can be sure of solving the problem of cancer within a very short time.

However, at this point, I have a particular view regarding cancer cells. In my opinion, the cosmic and ultra-cosmic radiations on Earth are causing a new kind of transformation on the course of life, of which cancer cells are one of its products and will replace all present cells within millions of years.

Also, because of the lack of favourable conditions for their life, cancer cells are currently a symbol of a type of disease, and this is not a true premise. I suppose that in the future by discovering the existence and effects of new kinds of radiations, using microscopes and electronic means, it would be possible to create a more peaceful yet temporary environment in which different body cells could live.

The difference between the structure of a non-cancerous cell and a cancerous one is in the volume and rate of division during the cell cycle. The relative time taken to complete the cancer cell cycle is two and a half times faster than that of an ordinary cell.

If the aforementioned theory becomes certain and acceptable, then accordingly, we have to believe that, during the next few billion years, the average human life expectancy will be much lower than it is today. Furthermore, due to rigorous life activities, the human being's power of thought will be many times greater than at present.

In any case, what we will witness by the year 2000*, as a result of the efforts of cell biologists - and I have discovered this through inner vision - is as follows: Human tendons and limbs will be constructed using suitable means (other than plastic surgery or transplantation) to be used in the body**. This will also be feasible for different blood types, and soon humankind will be successful in synthesising blood even from plants*** or other materials to be used when required. The only important subject - the sole way to these above mentioned achievements - is the experimental investigation and subsequent discovery of all kinds of radiations and energies in our environment, as well as their relationship with the chemicals and cells of our body.

Since galaxies are the consequences of explosions in infinite space, which after cooling manifested in the form of stars and planets, the nebulae and the energies which exist within all their atomic structures present or enforce their power in the same ratio.

---

* It is worthwhile to remember that *The Hidden Angles of Life was written in 1972 and first published in 1974.*

** Scientists have used stem cells to successfully repair and construct several complex tissues such as trachea, oesophagus and skeletal muscle in animal models and human beings. SF Badylak, DJ Weiss, A Caplan, P Macchiarini, "Engineered whole organs and complex tissues". *The Lancet*. **379**, 943-952 (2012).

*** Plants have been genetically modified to produce human blood proteins and tissue growth agents that could eventually be used to treat haemophiliacs. J Gao, BS Hooker and DB Anderson, "Expression of Functional Human Coagulation Factor XIII A-domain in Plant Cell Suspensions and Whole Plants". *Protein Expression and Purification* **37**, 89-96 (2004).

I want to conclude from this subject that when Earth settled in a balanced cosmic orbit, and presented life as it currently flows, its electrical capacitors were on the same level, and were connected with all the infinite energies. Earth is thus constantly under the influence of these energy exchanges.

The cause of the existence of beings, and the present manifestations of life, is the equilibrium achieved by the two forces of 'being' and 'non-being'. If one of these forces prevails over the other, or acts alone, it will alter the current forms of manifestations of life.

In Manifestations of Thought, which I wrote in 1954, I have explained death and birth in articles 7 and 8. Since they are related to these subjects and, to some extent, the spiritual aspects have also been considered, I have provided an excerpt as follows:

> Death is an evolutionary and developmental stage of being. It is the tearing away of the veil that covers the Absolute. Death and life are two reciprocal appearances that become thought of as stages of Existence. In reality, however, Existence is an absolute whole which contains these steps.
>
> Birth and death are clear, natural and precise images of change that follow an evolutionary pattern. Predestined, they move towards the freedom that has already been prepared in the universe. In the opinion

of physical and metaphysical scientists, the foundations of birth and death extend through physiology and anatomy to psychology and ethics. The Absolute Spirit stimulates evolution in the world of possibilities and presents many appearances, harmonising the conditions of its being. Before entering life in this world, it has ascended through many levels of ability and capacity on the evolutionary ladder. Following the same principle, it follows and accepts the equipment of the body and the organisation of physical life to continue its natural life. Thus, after death, it must neutralise the effects of the previous conditions, and move forward in the continuous journey, obtaining new equipment to harmonise with the environment. It follows this evolutionary pattern until it reaches a state of absolute freedom.

The multiplication of green algae and the reproduction of some types of bacteria represent this process of evolution which has been carefully studied. Yet in our limited understanding of the journey of particles and energy throughout natural existence, the wondrous mystery remains. We do see again and again, however, the general principle that perfection comes gradually and works precisely as it leads nature in the desired direction. Perfection gradually and precisely guides nature through the transcending changes, although these gradual changes cannot all be seen by the careful eye of the researcher. What we can see is nevertheless astounding, as we witness that Existence is always held in a delicate balance between life and death. One observable example of this gradual and immeasurable change is that of the yearly shedding of a snake. Although it is not precisely observable, the moulting follows a prescribed order. Similarly, a pan of water on the stove heats up slowly until it starts to boil; the changes take place gradually, but in the eyes of an observer, it happens suddenly, within a small amount of time. This same progression of endless sets of transformations happens every day, everywhere in existence. There is nothing that arrives suddenly, out of nowhere, or disappears in an instant. Instead, there are endless comings and goings, births and deaths, although it may appear to us that these are sudden events.

If one distinguished any pattern in the phenomenon of natural existence regarding the truth of identity, he would understand the essence of life after death. These different states of being and non-being are like isthmuses with a common boundary. Although they seem to be separate intervals, they do not have distinct and separate boundaries. Like the spectrum of light described by Newton, the different colours can be differentiated by the observer. Each colour has different specifications; red is distinguished from green and has different wavelengths. These wavelengths have all been measured from 1/1000 of an angstrom to 300 microns. Yet in a natural reflection, as observed from a prism, there is no distinct boundary between these colours. Gradually, red becomes orange, yellow, amber, green, blue, violet and ultraviolet. We can only approximately represent the spectrum. Although the colours have different appearances, there are no distinct boundaries between them. Similarly, transitional states contrast life and death, even though life and death are born of each other. They are continuous states of Existence. The unity and order which prevails throughout nature are aspects of the fundamental law of Existence. Continuity is invariably present in everything.

Thus, the materialistic view that matter dominates nature, especially regarding human existence, must be questioned. Hegel wrote that human beings are temporal states of eternal essence, evolution and potential. But if we compare our 'temporary life' with eternal time and space, we become aware of our nothingness, our existential condition.

Materialistic thought is based on hypothesis drawn from conclusions about natural, limited, observable changes. This view cannot and will not recognise the principal of existential originality through the developmental process and order of assimilated natural changes. In fact, the effective way of researching complex problems is not to merely accept what is observable and deny anything our senses cannot define. It is unjust to placate our minds with simplistic conclusions.

The natural and material structures, after undergoing transformations, are replaced by other limited manifestations. Even in the case of certain cells, such as the brain cells and neurons, whose slow rate of change misleads some people into thinking that they are eternal - or at least that they will remain relatively unchanged throughout their life to regulate and record the memories of life - it is undoubted that their manifestations and different forms have been ceaselessly changing over time before their birth, during their current life and after death. Therefore, no parts or organs of the body are the cause of the activity of human identity.

How could a constant identity be the result of continuous transformations and changes? In my opinion, the reason for this lack of recognition is that identity has always been considered proximate to matter and our observations have also been confined to the sensory activities of the body. We have therefore considered those same organs and parts – which are nothing more than a means to obey identity – to be part of identity, or even identity itself. Moreover, we have named sensory observations and experiences 'true knowledge'.

Real knowledge and cognition, through the essence of the soul, is dependent on knowledge of Infinite Existence, while personal knowledge and cognition tends towards knowledge in the limitations of beings, effects and surfaces.

Inward and conscionable research into knowledge and the metaphysical aspect of Existence, which is the final cause of beings, will be more useful than merely studying the surfaces of beings. Therefore, that which knowledge is seeking is the final and ultimate cause, i.e. the truth of life. The sensory sciences, in any form, cannot be able to penetrate realities. The human being is an intricate yet organised organism which, under the supervision and coordination of Existence, is, in the course of its life, engaged with accepting stability or changes within its own being.

Therefore, scientists and scholars in this era should not repeat the mistakes of our ancestors; that is, they must ensure that physics and metaphysics are not defined in a way that the current gap between them is increased. This is because, just as human society needs an understanding of physical matter, it also needs to understand morals and ethics. Apart from this, the structural limitations of matter, which imprison us in a superficial environment, cannot provide humankind with a vaster world.

In analysing the ideas of these scholars, we encounter this point that one's character and identity are constantly changing and never remain at a stable point during one's lifetime. This is because sensory observations and experiences continuously occur, and at any stage have specific characteristics which are completely different and distinguished from the past; and because the human memory and mind are obligated to collect those characteristics. Therefore, the character of the human being is continuously changing or growing and will be different from its past. Does this mean, for example, that a 24-hour-old baby, who as yet has no sensory understanding, has no character? So who is it then who cries when it needs food or is distressed, if it is not 'I'?

Suppose a person loses his memory in an accident, and later remembers those lost experiences after another accident. When such a person loses his memory, he does not show any irregularities in his actions. Yet, much like a newborn infant, he does not recognise objects and does not know their usage. However, he is curious to know these objects and to use them appropriately. Indeed, this curiosity and quest to investigate and cognise objects is proof of that innate knowledge whose facilities are not provided by an external source, but rather by a deep inner driving force, which is the identity of the patient.

Suppose it takes the patient one year to recover. Over this time he continually gathers new sensory and natural experiences. In fact, much like infants, he will examine those initial experiences and, by 'association of ideas', will decide whether or not to carry them out or dismiss them. Be this as it may, as soon as his cerebral cells become healthy once more and regain cellular communications, he will forget and discharge the recently gathered memories and sensations and, like a person woken from sleep, will readily recall past experiences.

How are these different identities perceived from that patient's point of view? Does this 'I' not know the former 'I'? Or could it be that during all these times perhaps there only ever was one 'I' whose apparent and mental impressions encountered a temporary coma?

By accepting and discharging these sensory experiences and acquired knowledge, the authentic identity proves that it is a knowledgeable Gnostic identity. Thus, with knowledge it recognises different types of experiences and knows which of its existing tools to use, and how to choose the right equipment for its organs to establish tranquillity in its being.

In order to perform their designated cellular programmes and continue to live, the human cells, including the blood cells, do not follow the will of the human being nor seek its assistance. Instead they do what is appropriate according to the universal order of Existence.

For example, if a microbe entered our body and commenced its activity, the body's system would start to repel it without notifying the human will or the body's other regulators, which have yet to be discovered. After a few days of continuous battle between the immune system and the successive generations of infecting microbes, the body's temperature would gradually start to rise above 37°C and only then would the carnal soul of the human being become aware of the body's illness and discomfort. Nevertheless, the white blood cells had already begun their calculated and excellent defensive attack against the invading microbes many hours earlier and had prepared for further defence by the proliferation of relevant antibodies. All of these important organised actions continue without the interference of human will, until the infection has been completely eradicated.

Certainly, all human cells are similar in their continuation of life, and their natural and material functions cannot have spiritual or qualitative products. However, there are organisations more subtle than the dense body related to the carnal soul and human magnetism, whose qualities form the basis of the dividing line between particles, waves and the Absolute Spirit. They rule through the intermediary stages and phases of their existence, and the extent of their manifestations in nature is based on their predestined connection and affection. Therefore, events cannot be the 'organiser' or the 'director' of the Eternal, but it is the existence of the Eternal which instantly determines the nature of the boundaries of events and instructs those events as to the conditions and circumstances.

In any case, these successive changes of the changing particles and waves cannot produce a stable product such as the unchangeable identity. Furthermore, the acceptance of such an idea does not seem to be logical either. Upon careful reflection on my natural life, I have noticed both evident and hidden changes happening in the 'life-course' of my body. I realised that there were so many things that I did not know but learnt of later, and learnt so many things that I subsequently forgot since they were of no use to me.

There was a time when the canvas of my mind was blank of any experience or feelings, and now many days and years have passed, and I have seen, heard and learnt many things, yet 'I' am the same as 'I' was; and with all these changes I have not witnessed any changes in my identity. It was 'I' who had the will to do something and it was 'I' who was deterred by them. In the end, it was 'I' who matched external observations with the contents of memory and created formulae for rejecting and accepting them, and it is 'I' who, with the power of thought and from the contents of 'the book of memory', create and visualise whatever I want.

I have found ability and knowledge in the vast Existence. This 'I' is not limited to a body for it to be cognised from the limited effects of that body - a variable and unstable body at that. So this 'I' who had remained genuine and unchanged in two different states, will, according to the rule of stability, remain genuine and unchanged in all other analogous states.

In his book Jnana-Yoga* Vivekananda, explains that each human being consists of body, soul and the real essence, i.e. the human identity (which is the same as that spirit). This identity is situated behind the body and soul. Body is the outer cover and soul is the inner cover of the spirit. This is a true identifier, end result and feeling of the body; which leaves its effect through the inner organs, or the soul in the body. The spirit is the only non-material essence, and because it is not material it cannot be a compound, and because it is not a compound it does not follow the principle of cause and effect. Hence, the spirit is eternal and because it is eternal it neither has a beginning nor an end.

Now, I have no reason to believe that my stable disposition (*Fetrat*) and the existence of my stable essence, which throughout my entire life has not accepted change and evolution, is nil and void before birth and after death. The reason for this is that the law which my identity lays down does not accept flaw; and that law is: A being, which does not change in two different times is genuine and immortal.

---

* In the Sanskrit language *Jnana* means "knowledge" and so Jnana Yoga is "the path of knowledge" (or "the path of cognition").

The character of an individual is, in reality, a stable soul and certain habits are transferred through genes, whose magnetic effects are within the extent of the brain's capability. With all these descriptions, the subject raised in question 11 would be correct if we could be certain that qualitative affairs, namely movements and motives, were individually efficient without any external relations. Moreover, genes are one of the particularities of cellular life. However, other guiding agents are the subjects of alternative visible and invisible forces, which have a separate section in Sufism, and I will explain a part of them, and their effects, in this article.

Two subjects have a direct influence on cellular life:

1. Nutrition
2. Temperature.

These two factors have formed the basis of the experiences of scientists, whether from primordial volcanic rocks or from present-age topics relating to the life, or birth and death, of the cell. For this reason we have defined life as balance in conditions, and change as disorder in conditions. Despite this we are heedless of the great effect of the cosmic rays, the solar, planar, and all of the other gravitational fields, both on the very existence of the cell and its present cycle.

Therefore, it appears to me that instead of spending all of our time merely investigating the cellular structure, we should first put our effort into building an apparatus that could analyse cosmic and ultra-cosmic rays. Then, study the effect of each of these radiations separately on the cellular cycle and the living state. Through this method, I am certain we will find more effective ways to correct the mistakes of the cellular life cycle in its fight against cancer or other similar diseases.

We should pay more attention to the energy stored in the hydrogen (H) atom. This energy is a deposit that a simple atom has carried within itself during its cosmic journey, ever since it was formed. I am confident that we will find the origin of life in the H atom or its compounds.

Our cells are basically composed of four compounds of H atom. Under the influence of cosmic rays and their energy fields, the stored energy in these compounds is constantly being traded between the internal and the external part of the cell. This exchange of energy is the very agent of life on Earth. Consequently, without knowledge of this energy deposit and its relation to cosmic radiation, we will be unable to discover the origin of cellular life.

Another factor which has a direct influence on cellular life is the electromagnetic centres in the human body, of which there are thirteen. These centres are in direct connection with the electromagnetic fields of the universe. With complete cognition of these centres it is possible to discover one of the existing intermediary states between the material aspect of the human being and his true character.

To explain the meaning of this we must refer to this example: Just as in a power station where electric transmission is carried out by means of wire cables through substations, in the human body, too, these centres assume the role of the substation centres and convert acquired powers to magnetism and suitable conditions for the body. The subtle materials of the body, such as magnetism and cellular organelles, are both receivers and capable, and each one of them is both effective and affected, i.e. are both active and passive. Therefore, when the function of one of these centres, for whatever reason, becomes defective, its reaction will be directly mirrored on the neural and magnetic structures of the body and they cannot fully perform their communicative functions, and consequently the bodily functions will encounter defects.

The transferred and released energies in the neural fibres, and so on, whose discharge in the body originates from these centres, are active throughout the body and in particular from around and above the ribcage; and they will flow out of the body from those parts of the body where the neural fibres reach the surface of the skin, like the fingertips.

These centres, which are spread through thirteen different points of the body, are in constant connection with the electric currents of Existence. To clarify the matter, I will explain the characteristics of some of them:

The most important centre in the human body is located in the hollow area under the sternum and over the diaphragm. Its magnetic particles are directly fostered by sunlight and posses horizontal filaments, which approximately intersect the neural fibres. This centre is an agent for many vital aspects and qualities such as: desires, fears, superficial love and so on. At this point I see it necessary to give, in précis, a definition of love and its states and stages.

The word love in Arabic (*Eshgh*) is derived from the Arabic word for bindweed (*Ashagheh*), a plant that sticks to trees and establishes practically a half shared life with the living tree. This is the most appropriate name given to this pure essence of life.

When life is defined in a much more subtle and spiritual level, then love must be considered a word similar to life. Moreover, according to the property of its existence it can be said that: Love is an all-pervading electromagnetic force, which unites and connects all aspects of Existence, from the smallest particle to the whole infinite universe.

In one of my works, Chanteh – the Gnostic's Cosmos, which I complied in 1941 at the age of 24, I defined love at a deep spiritual level as Inner Intellect or Absolute Eternal Intellect, and named its possible forms of natural affinities as affection. For example, the subject of chromosomes and their binding together is also one of its manifestations.

My father, the patron activist of Knowledge and Wisdom, the Complete and Perfect Sufi (*Aref*), the Supreme and Corroborated Hazrat Pir, Molana Mir Ghotbeddin Mohammad Angha (Oveyssi), in the introduction to his book *Az Janin Ta Janan* (From Foetus to Paradise) has stated about love:

> As I started to write the word 'love' this point came to my mind that since the majority of the Materialists have remained distant from the soul of love, they call sexual instinct, which is the lowest level and the worldly cover and the supposed sediment of the essence of human affection, love. They set the uprising of carnal lust, which originates from psychosis and is equal to schizophrenia, in place of love. They have seen nothing from the burning fire of love except smoke and have not reached the heavenly qualities of love. They do not know that if in the world of Existence the attraction of love did not exist, not even a particle would have been created. All of Existence is of love, or Existence is all love, which manifests as much as the magnitude of purity of each visible manifestation and is radiant in accordance with any degree of enlightenment.

In question 8, one or two distinctions of the human and animal brain have been noted. Now, if human character is detached from its cellular structure, many characteristic qualities will unavoidably be eliminated. The heavenly love between teacher and pupil, father and son, master and disciple, and higher than this stage, the relationship which the Holy Spirit of the prophets has with the Infinite Being, or

Unconditional Existence, or God is of this category. We come across this in Jesus Christ's saying of "my Father in heaven".

Physical, Spiritual and Divine love, are as have been described. Recognition of this matter is possible by paying attention to this law: Any countenance has a reality and therefore any countenance will discover its secrets when the veils and intervals have been removed, and this is the main principle of the teachings of Sufism.

In the book Love and Fate *(Eshgh-o-Sarnevesht)* I have defined love and destiny as follows:

> Love consists of a penetrating electric and magnetic force* which joins together the entire infinite universe. This is a definition of material love, whose antithesis is hatred; just as evil is the antithesis of good, ugliness the antithesis of beauty, and pleasure and suffering the antitheses of each other.

This feeling, to a degree, determines the fate of a seemingly small but everlasting group who are enclosed within the three dimensions of mathematics.

However, for those who have wholly attained the faculty of their intellect, that is to say they are annihilated in love, even the fate of the universe is nothing but their will. In summary, love is a word and a

* The electric force emanates from magnetic centres, similar to the way power lines transfer electrical energy from power stations.

reality, which mostly has a similarity in its being with life and God. Love is either intellect and life, or the most influential functionary agent of these two. This is because the effects of love in all possible areas of research are accompanied by a superior organisation of life.

In any case, because love in the material world is considered as a form of motion, it must possess a luminous reality and a stable and genuine basis, which again is called love. It is for this reason that love is sometimes referred to as superficial and material, and sometimes as spiritual love. Gnostics and researchers have termed the pivotal point of the two poles of true love and superficial love as intellect. In my opinion love in all three phases is the First Intellect, Second Intellect and Third Intellect*, which collectively sets in place the uniformity of Existence.

At any rate, as mentioned earlier, the centre for the effects of love and material desires is the magnetic centre under the sternum, namely the solar plexus [Figure 1, page 150, number 2]. Love at the 'pivotal point' occurs when this centre is in connection with the magnetic centres of the brain. Sublime spiritual love and prophetic love is related to the inter-connection of the Source of Life in the heart,

* The rational soul of man is said to comprise of a practical and theoretical intellect. The former being concerned with the physical well being of the body and the latter affiliated with attaining knowledge and cognition. In _Kitab-al-Shifa (The Book of Healing)_, Ibn Sina (Avicenna) discusses the three stages of the theoretical intellect, on its path to cognition. Starting with the "potential intellect" (*istidad*) which is inherent in the human soul from birth, it firstly assembles intelligible thoughts (*bil-malaka*) and then actualises them (*bil-fi'l*), before reaching the acquired intellect (*mustafad*) by means of the Active Intellect (or Universal Intellect, or First Intellect). (Herbert A. Davidson, _Alfarabi, Avicenna, & Averroes, on Intelect, Their Cosmologies, Theories of the Active Intellect, & Theories of Human Intellect_, Oxford University Press, 1992)

the solar plexus and the magnetic centres of the brain and their united effects on all magnetic and nerve centres for achieving a vast harmony and logical balance. Mistakes and wrongdoings occur when each centre performs dispersed tasks separately. Human identity becomes universal when true union and balance exists between the external, internal, individual and universal forces.

Therefore, we must be certain that through laboratory research carried out on the sources and centres of particles and the universe, it will become possible to resolve many spiritual issues and the gateway to the realm of the Absolute, and Reality, will open.

If the research of the spiritual scholars had started from this point, we would have understood many aspects of life by now. In hindsight, perhaps those preliminaries were necessary for us to be able to draw accurate conclusions today.

Thus, a school must also be established whose most fundamental principles are based on experiences gained from knowledge of the structure of magnetic bodies and their relation to life.

In any case, with respect to the magnetic centres and their functions, we will analyse the subject in this way:

For maintaining the health of bodily organs and ensuring the regular functioning of cells, there must be a suitable equilibrium in the magnetic functions, in assimilation and repulsion, and in the body. It is possible for this to occur and become operational, either directly by vital radiations or indirectly by the effects of these radiations on other natural substances. This is a novel guide in the medical sciences and a revision of ideas concerning the psyche, neural diseases, and similar subjects, and finally the discovery of more certain principles in biology. In my opinion, with this proposal many untreatable illnesses will become curable.* With this description we must acknowledge the necessity of inventing or perfecting electronic devices that can accurately analyse cosmic rays.

Another centre is located at the hollow area in between the eyebrows** [Figure 1, page 150, number 10], which has a direct relationship with the cartilages of the nose. Gathering and concentration of magnetism in this centre - and accepting the impact and impression of that agent*** - leads to telepathy, mind reading and further matters pertaining to the sixth sense.

Another magnetic centre is situated at the last vertebra of the

---

* Hazrat Salaheddin Ali Nader Angha explains how traditional medication works by altering the pH of the blood (and cells), and as such controls the intensity of "cold fusion" in the body (a fact oblivious to current medicine). Furthermore, some diseases require a higher rate of fusion than provided by manipulating the blood pH level alone. For these diseases, radiation therapy would be more successful if instead of using rays of photons, beams of muons were used, since muons would increase "the rate of fusion" sufficiently enough to eliminate the disease, something which photons do not achieve. (Hazrat Salaheddin Ali Nader Angha, *Expansion & Contraction Within Being (Dahm)*, M.T.O. Shahmaghsoudi® Publications, 2001).

** The third eye.

*** Concentrating on one centre will also gather electromagnetism at the other centres. Telepathy occurs when each centre is harmonised at the same level.

spinal column [Figure 1, page 150, number 1]. Some thinkers believe this triangle-shaped cartilage is an effective agent in the existence of the human being and is one of the major centres of life. Its structure is such that it will not decompose after death. In any case, the living structure of its cells possess a complicated composition. This centre is the region for exerting powers and influence, and is also efficacious in genesis, reproduction and generation.

In ancient beliefs and religions, and even in Islam, a great deal of importance has been placed on this triangle-shaped centre. For example, in the teachings of Yoga the spinal column is said to have two nerve channels, referred to as *Pingalâ* and *Idâ*. In the middle of the spinal column there exists an empty, hollow canal termed *Sushumnâ*. At the end of the hollow canal is something referred to as 'Lotus of the Kundalini', which has a triangular form. Whenever this manifests, the hollow canal will inflame and from there will find its way to the different levels of intellect until a 'mysterious power' is reached.

According to this theory, and just as it can be seen in anatomy, the spinal column is composed of horizontally aligned vertebra, shaped like the number eight or like the mathematical symbol for infinity (∞), arranged on top of each other in a specific manner. If these 'infinities' are placed on top of one another the right side of it is called *Pingalâ*, and the left side *Idâ*, with *Sushumnâ* travelling through the centre of the hollow canal.

The Yogi concentrates on his spinal column, but this form of concentration is completely different in Sufism. This is because the Sufi performs concentration on the Source of Life in the heart, and the centre below the sternum and from there expands to the brain. This type of unification and method is exclusive to the Oveyssi School. As I have explained in the book Message from the Soul, this method is based on the teachings of the prophets, namely: ‘Presence of Heart’, which is in touch with a more certain spirituality. Therefore, the heart must be known as the place for the emergence of all spiritual cognitive illuminations and prophetic systems.

This explanation will correct the ideas and mistakes of those who surmise that the practical method of Sufism is similar to the methods of Yoga. For this reason some have conceived that, because of his unilateral spirituality, the teachings of Jesus* were adapted from the spiritual methods of Buddhists and Hindus. However, because the Bible recognises the heart as the heavenly point, the differences in the methods of worship become apparent.

Moreover, this concept of a direct connection between the Source of Life and the brain has permeated Islam more deeply. With regards to this, I have given an in-depth explanation in the book Message from the Soul, which is part of three treatises translated into English as The Mystery of Humanity:Tranquillity and Survival, of which I have previously sent you a copy.

* Author’s original footnote: I have used the Prophet Jesus as an example since you are more familiar with him, although all prophets of God have emphasised the same point.

Of course, as you are aware, in the various methods, old and new, the path of spiritual cognition is not restricted to this one premise. However, for further clarification of the concept and importance of these magnetic centres it was necessary to mention the basis of yogic methods.

The location of another centre is at the crown [Figure 1, page 150, number 11 & 12], which acts approximately like a radar and as a result can connect with released thought waves in space. I believe inventors and discoverers find successful solutions through concentration of magnetism in this centre and its connection with the centre at the hollow area in between the eyebrows. This centre is capable of transmitting the brain's magnetic waves up to and beyond 12 km in space in the form of a vertical column of light.

Once alpha waves make contact with this magnetic centre and establish a connection with the sub-centres of the various senses, they induce the hearing, smelling and seeing of invisible waves and the events derived from them. It has been established that the connection between the grey matter of the brain and the solar system spans over a diameter of around 3 billion kilometres, and that alpha waves and other thought waves continuously interact with this connection. I have called this region 'the region of illuminated thought'.

Another magnetic centre is located at the head of the medulla oblongata [Figure 1, page 150, number 9] and is directly in connection with the last vertebra of the spinal column. When these two pivots are in direct contact with each other and can influence the spinal cord, the human being will be able to perform extraordinary actions.

The magnetic centre of the heart, which I have named 'the Source of Life' [Figure 1, page 150, number 13], is recognised as the centre for the propagation of life and as a perfect, wise and powerful director, which is in connection with all the magnetic centres. It is very likely that this is the Divine Point of the human being, which in religions is referred to as the Heavenly realm.

With respect to this, some are of the opinion that when religion talks about the 'heart' it is referring to a certain qualitative aspect and is intended towards a specific centre, whereas this is not the case and this 'heart' is the same as that which resides in the chest. This issue is implicit in religions and is clearly emphasised in Islam.

The Holy *Quran* (22:46) states:

> Do they not travel through the land, so that their hearts (and minds) may thus learn wisdom and their ears may thus learn to hear? Truly it is not their eyes that are blind, but their hearts which are in their breasts.

Or in the Holy Bible, Matthew (13:24):

> Another parable put he forth unto them, saying, The Kingdom of Heaven is likened unto a man which sowed good seed in his field.

Or in the Holy Bible, Matthew (13:31-32):

> Another parable put he forth unto them, saying, The Kingdom of Heaven is like to a grain of mustard seed, which a man took, and sowed in his field.
>
> Which indeed is the least of all seeds: but when it is grown, it is the greatest among herbs, and becometh a tree, so that the birds of the air come and lodge in the branches thereof.

The Torah, Exodus (35:10):

> And every wise hearted among you shall come and make all that the Lord hath commanded;

Or in the Holy Bible, the Book of Psalms (26:2):

> Examine me, O Lord, and prove me; try my reins and my heart.

Or in the Holy *Quran* (42:24) it states:

> What! do they say, "he has forged a falsehood against Allah"? But if Allah willed, He could seal up thy heart and Allah blots out vanity, and proves the truth by His words. For He knows well the secrets of all hearts.

It should be known that these types of centres exist for each one of the five senses, whose functions and relations are known to a certain extent. Other types of weaker centres, which are the substations of force, can be found in the body such as the lymph nodes.

Time and space for electromagnetism is different to that used for basic physical and natural measurements. That is to say, due to their subtlety, the extent of electromagnetic time and space encompass vaster and more sensitive realms. Exercise and special practices strengthen these centres and each one of them is the source of separate phenomena.

In Hamaseh Hayat (The Epic of Life), a book of verses covering many fields, especially regarding the current and future science of humankind, as well as in my other written works, I have explained these topics as follows:

The centre of the heart's solar plexus
Fulfils the intermediary state of the heart and brain

There are thirteen magnetic 'stars'
Supporting and carrying life

Positive and negative auras are with you
The sixth sense from this became true

From your chest, head and back
It surrounds your body by the width of four fingers

'Tis a luminous pinnacle, a protective shield of goodly Spirits
'Tis a manifestation of sincere worship

The body's electrical current is from this force
Concentration of hearts is its role

This aura is the first intermediary phase of the physical body
Covered in this mystical current is the physical body

The function of this intermediary phase - the etheric body
Is to attract and repel, to accept and stimulate

'Tis neither a metaphor, nor assumption, nor paradigm
Its effect on the heart induces a response

Disconnecting this source from the muscle of the heart
Privation on the effects of life will impart

The flow of this source is in three orbits
Point by point, orbiting around this compass

Concentration and dispersion are one of its effects
Desires are one from thousands of its acts

Its light, 'tis possible to see
Its effect, will manifest in thee

If this force is nurtured
The brain will be illuminated by its power

Your knowledge will surpass the limits of dimensions
These movements will gain motion from you

These explanations make it completely clear that recognition of these structures, which are unknown to contemporary science, is essential. Moreover, they must be considered and studied as being united with the neural system of the body.

Therefore, I envisage that instead of solely following one type of ambiguous philosophy, modern psychology will eradicate many unknown diseases by way of creating harmony between the human beings' magnetisms and by establishing harmony between these experiences, physiological medicine, medical healthcare and physical therapy. This is the method which medical science, willingly or reluctantly, must pursue

In addition, because the function of these magnetic centres - which ultimately includes the body - is directly related to the neural fibres of all cells, including blood cells, and is dependent on receiving magnetic waves from the sun, treatment of many types of neurological and haematological diseases will be possible through solar energy.

As you are aware, there are currently many spiritual healings taking place through the intermediation of different individuals. This process involves transmitting magnetic powers to the patient. This means that the healer is mediating and promoting strength in the magnetic body of the intermediary exactly like a magnetic centre.

There are many examples of these acts in the field of spiritual cognition. As a case in point, and as a witness, I will include an event which incidentally happened to me whilst writing these articles relating to magnetic forces. Although there are many examples of such observations, I recall this event because it occurred at the time of writing this section of the article.

At 8:30 pm, on 13 July 1974, a gentleman [whom we will refer to as 'S'] was involved in a horrific accident on a road with a young motorcyclist [whom we will refer to as 'K']. The crash was so intense that the injured man, who received fractures and trauma to several points of the body, was immediately taken to a hospital's Accident and Emergency Unit and put under primary care. The next day, at 5pm, he was placed under direct supervision of physicians. According to the hospital reports, the patient went into a state of coma within an hour of being admitted.

Two of the hospital's orthopaedic surgeons, as well as a specialist neurosurgeon from America - who incidentally happened to be visiting Iran for one month during this time - stated that, in addition to the fractures, injuries to the ribcage had resulted in fat molecules entering the bloodstream and caused the patient to fall into a state of coma. According to this, and based on the surgeon's report, once the fat molecules reached the cerebral arteries of the victim, the end result would be a stroke and imminent death; and there was nothing they could do.

This state of coma lasted for approximately 30 hours. At 11:30 pm the next day, i.e. on 15 of July 1974, 'S' came to me asking for help with the aid of a member of my school. I asked him for the name and address of the hospital.

At 2:00 am the next day, just three hours after their visit, the patient came out of coma without intervention from physicians. According to the surgeon's report, the patient's state was rapidly improving and based on medical consensus his survival was unexpected.

I performed this through the effects of concentration on 'the Source of Life'. This is a type of distant healing that does not require the healer to be in contact with the patient.

Concentrating the life forces of the human being on each one of these centres gives rise to extraordinary effects the likes of which cannot, in general, be found in lower animals to the same extent as that which is seen in humans.

In the same manner, dispersion of the magnetic powers in cases related to the sixth sense appears to show the human being as less capable than some animals, even though this is definitely not the case.

For example, in some circumstances animals are aware of an earthquake one or two hours before it occurs, which might be due to their 'feeling' or 'animal instinct'. However, as a result of concentration of the internal and external forces, 'perfect human beings' are aware of many events before they occur. Discovery of future events, and the ability to hold past incidents in sight, depends on the ability of the observer to be firmly settled in the present moment.

In Sufism the case in point is referred to as 'breaking the boundaries of limitations in an infinite environment'. Another explanation of this subject would be to say that when human beings succeed in overcoming the speed of light, they will become aware of the past and future manifestations of life, and in my opinion will cognise life at the level of beings. As an example, in 1961 I was in the presence of my great teacher, my father, when he uttered:

"I hear cries and moaning and see many ruins, under which people are buried".

Exactly a week after this proclamation a destructive and horrible earthquake, with its epicentre 60 kilometres from Tehran, shook an area of approximately 1000 kilometres in diameter, causing massive devastation. Consequently, about ten thousand people lost their lives in the tragedy.

My students have frequently witnessed various matters of this type. This is a principle which is adhered to with complete attention in practical Sufism, and perhaps most religions have such instructions, which are beneficial for the person performing them.

At the first opportunity, I will proceed to write a book which completely explains the human magnetic body and its properties and characteristics. I anticipate that this subject will be useful and noteworthy for the medical sciences, and will instigate a change to treatments in the future. Just as different parts of the body, with their specific duties, are related to and have an effect on one another, these centres are also in connection with each other and affect one another. The extent of their effects, from the body to magnetic particles and vice versa, is continuous and they are endowed with numerous signs of life. What is more, the first stage in the physiological death of the human body can be related to malfunction of these energy transmitting systems.

It can be said that memory, which is an interesting topic for researchers, is one of the qualitative effects of the magnetic field of the brain. This means that for each of the five stages of memory (learning, storing, remembering, recognition, and replacing) the field of activity is the inner part of the neurons of the cerebrum, which causes changes to occur to the RNA molecules and as a result actions and reactions in the axons activates the synapses (the place of neuronal contact).

To clarify the function of memory, it may be better to consider these synapses as magnetic plates which are activated by the brain's electrical impulses. It is from the conglomeration of millions of these microscopic plates that a grand and mysterious magnetic field, known as memory, is created which can record and store thousands upon thousands of auditory and visual waves. Furthermore, when necessary any one of the contents of the field, like 'files' of colourful slides full of information and images, can be visualised – consciously or unconsciously. New and healthy cells are more sensitive to these interactions than fatigued and weaker cells.

When we analyse a material compound and consider the ratio of its primary constituent substances, we will have a better understanding of the cause of its existence, its composition and the effects of the actuality of its compound, with the reality it possesses. The human figure and its constituting elements are composed of two conducting factors and one protecting factor. Of the two conducting factors, one transmits and the other receives. The protecting factor, which shadows the 'real character' within human beings, has a slower life course than the other two factors. The informative, or active, groups within cells are also made of these factors. We should not assume that the human being is merely this material structure.

It can be seen that the majority of intellectuals consider the magnetic connections of the body as spiritual cognitive connections, even though the independent character of the human being, which directs all these structures whether internally or externally, is an inconceivable Absolute centrality known as the spirit. The spirit, which is an active intellectual agent in the human body, must have this general definition: Spirit is vast and extensive awareness, and the body, its lowest state, is its medium.

I frequently contemplated how our 'awareness' would be after natural death when we have lost the factors involved in thought such as brain cells, magnetic and neural interactions and ultimately memory and perceptions.

Of course, what I had previously acquired from the spiritual realm did not fully answer this question, until on Wednesday 16 April 1975 at 2:00 am, whilst being fully awake and sitting, perhaps I had a spiritual separation, or ascension. In any case, whilst in a spiritual and conscious state, I encountered the spiritual current of a deceased pious man. I asked him to explain and clarify some aspects of the afterlife for me.

He said: "Spiritual currents are capable of penetrating thought waves".

When I asked him for further explanation, I became aware of an invisible force separating from him and travelling through my mind and brain cells. It recorded the thought and magnetic waves of my brain, which were just about visible, and exited from the back of my head, and whilst being directly positioned in front of me, he started to recite my thoughts; just like a computer, he recounted all that he had gathered from the faculty of my mind. He then went on to say: "Thoughts are not the truth, unless your thoughts fuse* with your reality, such as I have" (perhaps he was referring to his own spiritual current).

He then said: "You are capable of doing this".

With complete concentration of the heart, I empowered that thought within myself until my spiritual current took the form of extremely vast waves. Of course at that moment I felt proud and jubilant within myself, like a successful champion, perhaps because my thoughts and my spiritual ascent were on the same path.

The reason why I say that this discovery occurred in an awakened and conscious state is because I got up to do something and I humbly asked him to wait for me until I came back; and this is how it happened. When I returned, I saw him in front of me and we continued with our questions and answers.

His final remark was: "Tuesday 17 June", and after that I did not see him anymore. I later checked the calendar; the date and day corresponded. Now I have to wait for that date. On that day, I will

---

* When sugar dissolves in water the essence of sugar, $C_6H_{12}O_6$, interacts with the essence of water, H_2O, and as such it becomes embedded within water. It can be said that sugar and water are unified. In the same way thoughts coming from the waves of Existence must be embedded within the core of your reality, and emanate from your reality for them to be "Truthful".

understand what he meant by that remark. Could it be that he wanted to verify the day of the week or date of the month? Or did he have another intention? (Personally, I feel a certain transformation within myself). There still remain two months from the date of writing this correspondence to that date.

If this experience appears, on the surface at least, to be somewhat similar to telepathy, I must bring your attention to this point that the two are by no means related, and it would be better if we considered the topic according to what he (that spirit) said: "Spiritual currents are capable of penetrating thought waves". I have not come across any such issues in the writings and works of spiritual scholars; and even for me this has been a novel case.

I envisage that this event could to some extent solve your problems concerning the phantom.

In order to be aware of the realm of the Absolute, it is necessary to cognise and follow the Source of Life in the heart, the centre of which is the divulging and effective agent of life in human beings. It is for this reason that in the works of the prophets less emphasis has been placed on the spiritual activities of the brain. Even the natural thoughts of human beings, which are the products of comparisons and the relations of the mental faculties, are considered to interfere with spiritual harmony and solitude of the heart. Therefore, the teachings of the prophets in general and Islam in particular, for journeying through the Heavenly realm and discovering the Absolute realms, are based on the centrality of the heart.

One of the distinctions of the heart, in relation to the human body, is that it nourishes and controls all the organs, but is itself solely responsible for its own control and nourishment. Furthermore, the Source of Life in the heart is the origin of the first and last signs of physical life in human beings. It can be said that: The spirit of the heart is the un-recognised character of the human being. With regard to this, the Holy *Quran* (50:37) states:

> Verily in this is a message for any that has a heart and understanding or who gives ear earnestly witnesses (the truth).

In the book Message from the Soul, and my other works, I have explained the distinguishing characteristics of the spirit of the heart, i.e. the real character of human beings, which is the same as the stable and infinite 'I'.

Natural scientists have differentiated cells and particles in accordance with their respective functions. In my opinion, human forces should also be distinguished and identified with respect to their physiological functions. I do not see it necessary to reach highly elevated spiritual levels in order to discover this goal. Instead we must involve the initial connections, which are these magnetic currents, within the extent of our research. The path to reaching higher, more subtle goals, whilst respecting order, will then be possible through numerous intermediary states.

Thus, I propose that in order to cognise life, or its mysteries, we should immediately deem the experiences gained from the laboratories studying matter as a preliminary and move on to studying the content of the Book of Creation from its initial chapters. Then we can discover what effects attraction and repulsion, human magnetisms and the planetary and cosmic forces have on particles and how they maintain the comprehensive balance of life.

As an example of magnetic visualisations, it is worth mentioning an event which happened simultaneously in the presence of eighty of my students in Tehran and fifty in Abadan.

At 10:20 pm, whilst teaching my students in Tehran - and whilst holding a *Tabarzin* [Figure 2, page 151] (which is a symbol of the Oveyssi School) in my hand - I found myself amongst my students in Abadan, some 1000 km away. That is to say, I was present in two places simultaneously. Once this period of disappearance and subsequent reappearance had elapsed, I announced its time to my students in Tehran. A couple of days later we were informed from Abadan that I had been there at the same time and that some of my students had heard me speaking to them and had in turn spoken to me. Currently those students are present in my school and are writing and publishing accounts of their observations as witnesses to this and other similar events.

How can science, with the assurance and clarity that it has, explain this event?

Therefore, by consideration of the above issues as well as the fact that a cell is, both internally and externally, under the influence of groups of unknown radiations and magnetisms, discovering cellular life without any knowledge and understanding of these forces will be impossible. I wish I could assure the scientific community that the gateway to the discovery of the origin of the cellular living state is understanding the individual and collective radiation fields of the cells and their interaction with all of the cosmic radiation.

If, in conjunction with the experience and research of the cellular organisation of the body, the electromagnetic figure of human beings presented its assured form under detailed microscopes (which in my opinion it will), then at that point the perspective of research would observe 'true life' from more sensitive scientific angles; and we remain in anticipation of those days.

Apart from the electromagnetic body, we should be certain of other intermediaries between the Absolute Spirit and the dense body, whose recognition and attainment through experience will certainly be possible in a stepwise manner. We must first be aware of the human electromagnetic body since its recognition has certain criteria for recognising more subtle intermediaries.

The human being has always thought of rescuing himself from the hardships of materialistic and spiritual misfortunes. He has searched for a remedy and compiled an easy, wise and beneficial path in the form of correct programmes, whilst bringing them to the level of belief and action.

In the works and beliefs of Buddhism it has been explained that if man recognises his own self, no longer will he desire to continue this earthly and physical life, because when the self attains that true cognition, desires and wishes will naturally disappear. Ignorance is the strongest bond keeping human beings interested in this world and its attachments. Knowledge and ignorance are two paths extending in opposite directions. He who chooses knowledge will not be misled by countless carnal pleasures and wants; and he who has chosen the dark path of ignorance will benefit from nothing but vagrancy.

The learned who has cognised himself neither feels sorrow nor joy. He is needless of too little and too much, of good and bad, or of present and future.

This does not mean that he will not rise to support or guide society; rather, it is goodness which is the means for the existence of such a broad-minded person. Moreover, he does not anticipate remuneration from anyone for his social aid. These individuals comprise the group of true prophets and gnostics, of whom an example is the Prophet Jesus.

Only knowledge and understanding of wrong and righteous deeds can enlighten the path of truth and righteousness and make us virtuous.

With the above explanation, the importance of the School of Islamic Sufism and its expansion throughout the world is discernable, because it is this very cognition and morality that distinguishes human beings from animals.

On the basis of results from geological research, including discovered fossils and the age of metals, scientists and researchers have presented taxonomical classifications and criteria. Hence, the age of the planet we live on is estimated to be 4.6 billion years, and just as it has been in the past, it is currently progressing towards an eternal perfection. However, due to a lack of self-confidence, most of its inhabitants are still scared to accept the truth.

They fear that they may lose those false references which, based on national traditions, were created according to the laws of inheritance or individual and social 'accepted reasoning' of an imaginary fear. They think that if this should happen they will be like a wondering sphere in the hands of fate and destiny, following an unknown destination which will ultimately lead to death and eternal nihility.

Throughout time, whenever the human being found himself to be insignificant and helpless in comparison to the natural forces, he would regard virtually all sources and effects of the manifestations of natural phenomena such as thunder, lightning, rain, earthquakes, darkness, the seas, the sun, the moon, the sky, fear, horror, kindness, love, peace, war and so on, as more powerful than himself. Inevitably, he decided to worship the will of nature in the form of deities with specific manifestations and to show his admiration for Jupiter, Venus, Mars, Neptune and Uranus by offering sacrifices and performing specific worshipping rituals. Moreover, to be immune against dreadful and inauspicious events, and attain his inner goals or his own wishes, and to carry out his own will, he would seek help from their presence; and to suppress their burning anger he would offer his own offspring by throwing them into the waves of the rough seas and into unruly flames. So that by the law of 'focused attention', belief and establishing a sole will and certainty, he would prosper from his own intentions. He would then accredit this to the abundant grace of his precious deities, which he himself had created and given character to.

Knowingly, or otherwise, the priests nourished and encouraged these fabricated thoughts, which contained a distant truth, and spread them amongst people to such an extent that we can still find remains and traces of this particular type of faith amongst some African tribes, such as the Hottentots and Bantu, or amongst ignorant members of any society.

Primitive man's ignorance of the natural forces, their manifestations, and the fear which overtook him when they occurred, made it a necessity for him to respect and worship different kinds of deities in every era as his most valuable protectors. Some groups following these faiths then added branches and spread the beliefs. Be this as it may, had the deity of 'faith' not fulfilled man's expectations in accordance with his personal beliefs and true individual merit, and had it not answered his prayers to the extent he wanted, certainly an assortment of other deities would never have been worshipped.

However, less respect and attention has been given to this honourable deity who was the gatekeeper of cognition because, unlike other idols, it did not have a certain appearance or specific form and as such did not have a place in the temples. The truth is that the strength and ability of this powerful deity is far greater than that which it manifests and whatever has emanated from it has been accredited to other deities.

Even in the presence of all this, the deep influential voice of consciousness and the true face of human disposition (*Fetrat*), which was hidden behind these strange masks and made-up idols, never stopped its constant manifestation. At every moment it notified humanity through a speaker's influential voice.

Both savagery and civilisation have heard the voice of truth from nature's eminent speakers. In any case, the purer souls, meaning those with the talent to receive guidance, have relied and hoped upon these signs, even if they bowed down to or worshipped an idol or a tree. Ultimately, had these signs been present in the form of stony idols, or had the superior and trained souls attained the real guardianship of Existence, they would have been the models and leaders of human civilisation.

However, that exposed truth and untainted Absolute identity, which from the very beginning entrusted Unconditional Existence to every being, has become an abode for the selfish greed of society to such an extent that a novice and naive seeker of truth cannot easily break down and analyse this from the many forged veils and void counsels which have become generalised.

According to Xenophanes if oxen and horses or lions had hands, and could paint with their own hands and like humans, were able to produce fine arts and handicraft, then horses would surely paint gods similar to horses, and oxen would paint gods like oxen, and each would give their gods figures and bodies like themselves.

In any case, these mistakes, which cause an internal and spiritual restlessness, prevent novice scholars from cognising or showing the slightest interest in discovering and recognising the greatest secret of nature - their own self. For this reason they are consistently content with and worship an imaginary, far-fetched, non-existent goal and a non-settled, unstable, vague and unreal refuge. In the occurrence of any events, and to fulfil selfish needs, they disregard any perception of reality and seek help from - and are satisfied by - that unlimited unknown which has thrown a veil of darkness over their conscience.

Spinoza has said that the more a self knows its powers, the easier it can guide and organise itself; and the more it cognises nature's discipline, the easier it can detach itself from useless matters.*

On the basis of what has been explained, we have to confirm that it is the spread of baseless illusions and superstitions, which keeps the inexperienced yet relatively curious and eager minds wandering without purpose for some time.

Through self-research, an individual will, at the very least realise simple and fundamental principles capable of guiding him to inner goals. He will not be misled like those who merely observe sensory actions and occurrences without intention.

In fact, the human being is the 'computer' of the infinite universe. It acts according to the principles Existence has given it; regardless of whether they are true or false. In order for the mind not to make mistakes from the limited and sensory observations of natural life, the infinite universe is said to be like a constant 'film' onto which every detail and necessity is 'recorded'. Nothing is ever forgotten. Its 'playback' is exactly this, which the universe is in. Therefore, within

---

* In Part V (Of the Power of the Intellect, or the Human Freedom), Proposition 10 of his most famous work, *Ethics,* Spinoza states "As long as we are not assailed by emotions that are contrary to our nature, we have the power to arrange and associate affections of the body according to the order of the intellect". He further explains that this is possible so long as the "power of the mind" is not impeded in its "endeavours to understand things" and as such has "the ability to form clear and distinct ideas, deducing them from one another". (Benedictus de Spinoza, *The Essetial Spinoza: Ethics And Related Writings, edited by Michael L. Morgan; with translations of Samuel Shirley,* Hackett Publishing Company Inc., 2006).

the depths of the reality and laws which exist, we cannot believe that certain types of movements, changes or shortcomings must reach perfection.

These types of topics, which we have discussed here, have been considered in the highly advanced Sufism of Iran. The books *Chanteh: The Gnostic's Cosmos* and Manifestations of Thought, written by me, clearly prove these subject matters, because if we want to consider science as untrustworthy, then all the research conducted by scientists throughout the ages would be worthless and illogical. Since we may think that such a day will arrive, on that day a more transparent knowledge will manifest.

Movement is a type of agreement established by limitations, and time is a definition for measuring a broader limitation. These are the types of agreements which divert human thought, because the principles that are suitable for limitations are not in the same category as those necessary for the Infinite. By considering this explanation, if we wanted to design and build a computer, like a well-informed and pre-eminent scientist capable of solving our problems, we would have to use the guiding waves of human life at suitable levels. This, I am sure, will become possible for humankind, through controlling the various waves of Existence.

Would you then suppose that this knowledgeable computer, programmed by human beings to display such balanced logic, is a soulless apparatus? Or, it should be said that it possesses a universal and infinite soul, just as the human being in comparison with all other creatures possesses a holy and supreme spirit. Thus, the criterion for its uniqueness is not based on its own limited will. It is on that day that the human being will live entirely in the form of a computer.

In order to offer a more certain life to humanity we must gain access to the magnetic points of the cells, and their connection with their own environment and the world. In any case, besides the material elements of the cells and the active and passive powers of the world of particles we must also consider a kind of Absolute reality for every particle.

Let us suppose that an individual person is a cosmic cell: His organs would then be easily considered as the numerous and different systems, and visible and invisible faculties, all of which have a direct influence on his life. In addition to all this, there exists a personal conscience which no one is aware of except him. This personal conscience, which he calls **'I'**, is neither an animal conscience born of cellular actions and reactions, nor is its being reliant on the ever changing particles of material structure. However, it must be said that this is a balance of life, required for the manifestation of every being.

One of the main principles of Islam is balance; and balance means the stability of essence, not something opposing cruelty. This means that whenever there is real justice there is Existence.

Consequently, once we have cognised the real and complete structure of a cell, we must still be sure that the ultimate cause of its being, or its conscionable character, which is the responsive agent to all questions, actions and reactions, still remains hidden.

Is it possible to extract the reality of the human being from within the intricate structure of cellular particles by surgical techniques and use what has been obtained from that concealed place to describe human reality as it is?

The point is that for this reality, or any other truth, descriptions using names and experiences are veils which conceal the truths; and they are certainly unable to help scientific and practical research in cognising the truth.

What is it that has motivated us to describe and experience the intricate world so simply, and has led us to close the paths of cognition?

Do you realise how complicated and unknown the word '**I**' is, which everyone uses without any consideration to make themselves known to one another?

Who can cognise '**I**'?

Only '**I**'.

Can anyone other than '**I**' cognise '**I**'?

No, because anyone other than '**I**' is other than '**I**', and, as it is other than '**I**', how can it cognise '**I**'?

The word '**I**', which expresses the existence of every being for gauging itself and presenting to others, is synonymous with Existence.

The truth of an object's reality is within its own reality and because it is Absolute, it is simply out of reach for material things, and as such it cannot be cognised or measured with units which are based on the senses.

It is necessary to become Absolute and cognise the Infinite. This cannot happen unless the intermediary states between the Absolute and Cognition, which are more subtle when compared with dense matter and more dense in comparison with the subtle powers of the Absolute, are firstly recognised and retained in order to provide some criteria. These intermediary states start from the natural and individual magnetisms of human beings and their relations with the cosmic powers, the etheric doubles, the phantoms and astral bodies; and continue until the invisible powers and the infinite realms, which are the Heavenly realm (*Malakut*), the Sphere of Divine Power (*Jabarut*) and the Luminous Kingdom (*Lahut*).

From a conscionable point of view, the stable '**I**' in every being is a reliable unit for measuring the world of spirituality and morality. If human existence continues its life journey in a balanced manner, benefiting from all that Existence offers accordingly, '**I**' will be capable of performing much deeper, and almost supernatural, actions.

The name of the Holy religion of Islam is derived from this real submission (*Tasleem*) of an individual to the Infinite. The effects of this real submission include:

- Foreseeing events before they happen
- Communicating with the spiritual character of others from afar or near
- Numerous visualisations using a variety of intermediary magnetic currents
- Healing the ill
- Awareness of surroundings, and of the manifestations of creatures, without any time restrictions; and even awareness of the realm of the Absolute before leaving the bodies
- And other similar events

I will describe one or two instances which happened to me and were witnessed by others. In 1966, in a very short period of time the duration of which I cannot specify - but to give an approximate timescale I would have to say that perhaps it took as long as drinking a cup of coffee - I felt situated in a cognising spiritual state. A wave expanded from my chest and surrounded a vast area beyond the range of the senses. Although my body did not appear to have moved from its place I was everywhere in that area and felt everything well enough to study them accurately. Within this short period of time, I witnessed the chronicles of future eras, as well as many planetary events, and the scientific course of humankind for half a century or a little more.

I want to explain this event in such a way that it does not lose its real sense.

In that state it occurred to me that a vast, direct and all-encompassing connection had established between the invisible forces of my body and the natural waves. Obviously, at that moment, I was either unaware of my physical body or my existence had expanded as far as my attained observation. I saw myself in other planets and later times, and was informed of what happened there. This reminds me of Flammarion's sayings in the book God in Nature, and even though the reality which I perceived may be somewhat different from his opinion, in some way they are similar.

At that time, I wrote a part of what my reality was able to attain within that period of expansion of my chest, and perhaps many others which I cannot recall now, in a book entitled Hamaseh Hayat (The Epic of Life), which has been published. The following is an extract from that book:

This universe, through its cosmic course
Will cease to progress without occurrences

All particles, from the planets to the cosmos
Continue on a journey without self will

Each moment within every era brings forth
A new appearance to the state of an orbit

Within such an Infinite and boundless Existence
The rotation of firmaments will not come to an end

Just like a particle which has an orbit
Wave upon wave will exhibit this manner

A wave, on its path, is led by another wave
Being is on a journey within its own existence

The rotation of firmaments is far from its destination
The cosmos is compelled to its own rotation

Far we say, but separation we do not mean
All within all is not separated

Just as a fish, from water it comes, in water it lives
And every moment, seeking water in haste it is

Of these manifestations that it performs competently
There is none that is done, unknowingly

Ever new happenings, Existence has within
*"Every moment in action is He"**

All this, more and less, from life and death
Is the veil over the appearance of the attributes from the Essence

How unaware are you of the Kingdom of God?
Glance towards the Celestial heavens

From your own heart make a path to the universe
Make your Soul aware through concentration within the heart

Free from the Ka'bah, Temple and Monastery
Journey to the Celestial heavens

Open the eye of the heart upon the horizon
Why does every particle yearn for the other?

All these suns, moons, and stars
Where to, and for what, are they rushing?

So that, not for a moment will they stop rotating
Rushing, leaping and interacting with acceleration

The night sky sparkling like a broken mirror
Open your eyes and behold the splendour of God

---

* Reference to the Holy *Quran* (55:29).

What are the fixed stars and planets for?
What purpose is there in this journey of the globes?

Venus is joyful in building itself
Mars is embroiled in a losing battle within itself

Venus looks towards the land of Eternity
Mars has lost hope from Existence

Venus journeying in its first era
Mars is bound to a convent from an old era

In another galaxy, there is one like Earth
But not in the same order, orbit or state

From the vastness of this confirmed creation,
Who saw anything except gathering and dispersion?

Each firmament has its own orbit for sure,
Containing the Sun, Planets and the Earth

The deviation that is hidden in a particle
Consider it, as the astral plane of Plato

Part and whole are the mesh for life and death
Proceeding drunkenly in the tavern of the universe

Two lives and two deaths there are from Essence
The book of Genesis for this life and death

The definition of creation of the universe is:
"motion" from "motion" in "motion"

Any appearance that manifests with time
Comes from the secret house of mysteries

The story book of the universe of creations
Is wrapped between two nihilities

In the horizon of Existence, and with every inspiration/expiration,
The oscillation of a wave draws the sketch of manifestation

An occurrence will happen from the Eternal Kingdom
*"Alif" from "Lam" will turn into the circle of "Mim"**

Existence of the particle will blossom
From nihility, the wave of Eternality will come alive

A particle that becomes a world creator
Will inevitably become an angle for secrets

---

* Reference to the Holy *Quran* (2:1). Symbolically, the written Arabic figure of "Alif Lam Mim" represents the growth and unification of the life force which is hidden within every being. (Hazrat Salaheddin Ali Nader Angha, Sufism: Lecture Series, MTO Publications, 2011).

Once the circle of the particle is borne of light
The universe celebrates anew

It is obvious that in the next century
A veil will be removed from the ignorance of humanity

The Knowledge of religion will become the fruit of vision
The Knowledge of the physical will all be renewed

These two will be facing the limits of the Heavenly realm
The Material world will take on the colour of the Luminous Kingdom

When the Knowledge of Light prevails
The Spirit of Sufism will be the talk of the day

Vision will have an enlarged field
The symbol of ignorance will change

The Principle of Islam will become global
After another century by the will of destiny

Earth will be enlightened from this Light
Creation will have another appearance

They will discover the colourlessness of the unknown
Cause and effect will reconcile

The footprint of your past days will be found
Your future image will reveal all

At an instant, you will cover a journey of two hundred years
As you will live in that halo

Distorted is what space dimension will be
Expanded is what the dimension of time will be

The mystery of converting substrate factors into one another
Will be poured onto the human's world by Knowledge

The secret of Being will be witnessed
From behind the veil of this being and appearance

Living entity uniquely will surface from non-entity
Due to changes in effects and properties

There are so many revelations behind the veil
Whose inner sight will become awakened?

There are one hundred and ten colours from a "Colourless" wave*
That displays the extent of the particle's capacity

---

* Hazrat Shah Maghsoud Sadegh Angha first used this term to refer to the energy distribution inside the penetration sphere of a particle, the form of which cannot be the same as that outside the sphere. (Professor Sadegh Angha and Dr. Sirus Aryainejad, *Professor Sadegh Angha's Theory of Particle Structure and Its Application*, Vantage Press, New York 1994).

Full to the brim from the waves of Life is Existence
They inscribe the circle of being and appearance

From so many continuous means and tools
The waves of Life are forever rising

Vibrations will arise from the depths of directions
*The secret of life shall be discovered from water**

Seven elements will bring you seven hopes
Three from the cosmos and four from the sun

White, yellow and red are their traces
Blue-black in colour is their rock

It is water falling onto fire
Silencing its struggle

Meaning that, due to cosmic strife
The soil will open its wings of purity

Wavelengths of this carrier will make complete
Motionless water through constant heat

* Hazrat Salaheddin Ali Nader Angha explains that the O_2 rich environment in the "suds" of the sea water provide the suitable conditions for the production of muons and subsequent catalysis of cold fusion, which is "the main and the most influential parameter in the appearance of life on the surface of the earth. In fact, life energy is nothing but fusion energy. Fusion is the very secret of life on the surface of the earth." (Hazrat Salaheddin Ali Nader Angha, Expansion & Contraction Within Being (Dahm), M.T.O. Shahmaghsoudi® Publications, 2001).

The world will begin a journey of perfection
Through the outcome of the journey of Gnostics

The science of conversion through this action and reaction
Will turn soil to wind and water to fire

From One everything will come to exist
Journeying will become timeless

Your nerves and organs, they shall make from the Earth
By the request of nature through deep Intellect

Artificial blood they will synthesise
With embryos, they shall do whatever they desire

You will come into existence, in every century
And, amongst all beings, will exceed

But alas the conflict of yellow and white
Will burn this green fertile land of hope

The Knowledge of the Qur'an will be the manifesting Light
Through guidance from the Exalted Lord

As long as the world exists, Mohammad (pbuh) is its guide
His wisdom is the charge for freedom

Our cluster is all from His harvest
The Knowledge of Existence is a trace from His treasure

It is as though in this 'journey of possibilities', I have considered clairvoyance similar to secrets and mysteries shown by the prophets; but with this difference that the range of their spiritual insight was deeper and more vast than the range of my vision in this event.

By encountering this event, and similar ones which have occasionally occurred, I have come to believe that the human being's electromagnetism is connected to the whole electric system of the universe, and that the neural system and the magnetic particles collectively form an amalgamation which governs the 'life' encountered by an individual.

It has sometimes happened to me that I have touched the magnetic rays of my own bodily organs or other external currents with my outer senses. I urge scholars to consider the scientific aspect of Tamarkoz*, as practised in Sufism, more diligently. It is because I am certain that human electromagnetism is on the same plane as subtle particles which are the means of life in the world of particles, and are sometimes active, sometimes passive or both together.

Suppose you position a lamp so that it is surrounded by mirrors in

* Tamarkoz® is the ancient Sufi art and science of self-knowledge through concentration and meditation. This method of practice is exclusive to M.T.O. Shahmaghsoudi – School of Islamic Sufism®.

six planes, in this case the rate at which reflected light bounces back and forth will be such that the reflected rays will be beyond the limits of mathematical calculations.

This example is not dissimilar to the status of objects in the world of waves. Therefore, the appearance of living and the form of life can clearly be understood. So all effects from particles, objects and the world which leave an impression on our senses are reflections from dimensional mirrors in the world of life; which come to exist through the acceptance and influence of events, and the invisible forces of the Infinite.

I state that the basis of all colours is 'Colourless' and that the reflection of light, with various descriptions, makes different colours visible and brings about a definition for the existence of beings, which might be completely different to what it actually is. Humankind, in general, considers whatever is felt by the senses to be 'Existence' and assumes that whatever is invisible and cannot be sensed, or at least for now is invisible to him and does not conform to sensory gauges, is 'nihility'. It remains heedless of cognising the fundamental principles of Existence.

Now, consider this scenario in relation to all creatures and the infinite universe. See whether this law of life: 'All beings are together and are interlaced' and ultimately a definition for unity which does not describe separation and connection and does not accept union and dispersion, but rather accepts continuity and a recurring cycle is correct.

Therefore, we would not be wrong in saying that the conscionable 'I' is the centre of the Infinite Existence and that any knowledge in its centre has conscious realisation. Thus, by cognising the reality of 'I', the great book of Existence becomes comprehensible.

Moreover, we must accept this law that:

'Logical thoughts exist'.

It is for this reason that in Islam it has been said that cognition of reality is confined to the cognition of 'I', and this 'I' is the most real manifestation of the Infinite Existence.

Hazrat Ali, that Eternal Man, that soaring Falcon of the Divine Throne, that Secret of Existence and Guardianship - thousands of greetings and peace be upon Him - who was one of the first teachers of Islam and to whom our school of Sufism is affiliated and from whom it has propagated, stated fourteen centuries ago that:

"You assume that you are a microcosm while the macrocosm is within you, and you are the clear book of Existence".

In the teachings of Sufism it is said that if all human forces are gathered at one point and at that moment no other subject or object is thought about, the accumulation of one's resulting forces is the essence of the generosity bestowed upon him by the universe, or Existence, or God, or whatever we call it.

The cognition that you can achieve in relation to your own being, the universe, or the particles is no more than the faculty of perception that you possess. This means it is not greater than the knowledge which has been granted to you. So it should not be said that the being of 'I' is anything other than Existence, since in this case we would disconnect all associations with Existence. Nor should it be said that the being of 'I' is identical to Existence, since 'being' is one definition for limitation and singularity, which separates it from Infinity.

With regard to this, the Holy *Quran* (15:16-30) informs us of the subject of creation and genesis, from the true viewpoint of the pious as follows:

> We have set constellations in the heavens and made them pleasing to the beholders,
> And guarded them from every stoned satan.
> Except for he who steals the listening and is then pursued by a visible flame.
> We have spread out the earth and set upon it firm mountains. everything we have caused to grow therein is justly weighed;
> And there appointed for you is a livelihood, and for those you do not provide.

> And there is not a thing but with us are its treasuries, and we do not send it down except in a known measure.
> We send the winds fertilizing, and we send down out of heaven water, from which you drink and you are not its treasurers.
> And surely, it is we who give life and make to die. We are the inheritor.
> We know those of you who press forward, and we know the laggards,
> And it is your lord who will gather them. He is wise, knowing.
> We created mankind from clay, molded from mud,
> And before him we created the jinn from smokeless fire.
> When your lord said to the angels: 'see, I am creating a mortal from clay of molded mud.
> When Ii have shaped him and ran my created soul in him fall down prostrating towards him'.
> All the angels prostrated themselves,

These verses are describing the face of creation. However, with regard to the reality of the Absolute, Maurice Maeterlinck has said it is better to say everything has always been and will continue to be, provided that we do not stop researching and exploring; even though our research is similar to a drill with which we are continuously deepening the well of our ignorance.* Despite all of this we have no other way ahead for discovering the secrets of the universe and life.

---

* In _Wisdom and Destiny_, the 20th Century Belgian Symbolist poet, playwright, and essayist, Maurice Maeterlinck, writes: "The sage is consoled and fortified by everything that exists, for indeed it is for the essence of wisdom to seek out all that exists, and to admit it within its circle." (Maurice Maeterlinck, _Wisdom and Destiny_, Dodd, Mead and Company, 1899)

This means that when powerful telescopes of the observatories expand the domain of science and knowledge by several million years, and weaken or nullify the majority of established scientific laws and conventions - for example, discovering that the movement of a red or white blood cell within our vascular system affects, or is affected by, the cosmos in accordance with the law of gravity, with a speed of one hundred thousand times faster than the speed of light - once again I will ask myself why it should be this way and, if it did not happen this way, what would have happened?

Even though the word 'why', which is used in all our research, may at first seem scientific, it is in reality a clear indication of our ignorance and incapability. Living in limitations seeks a particular purpose for every cause or event but, undoubtedly, for the Infinite Existence and whatever exists, these 'whys' do not exist.

In a translation of a section from the Rig Veda (153826),* dating back to 1400 BC, the German Orientalist, Max Muller states:

> Then there was not non-existent nor existent:
> there was no realm of air, no sky beyond it.
> What covered in, and where? and what gave shelter?
> Was water there, unfathomed depth of water?
>
> Death was not then, nor was there aught immortal:

* Friedrich Max Muller, *The Hymns of the Rig Veda, with Sayana's commentary*, London, 1849-75, 6 volumes; 2nd edition 4 volumes, Oxford, 1890-92. This passage refers to "The Song of Creation" (Book 10: Hymn 129), which is a section from the most celebrated translation of the "153826" Sanskrit words of the *Rig Veda* into English by Max Muller.

no sign was there, the day's and night's divider.
That one thing, breathless, breathed by its own nature
apart from it was nothing whatsoever.

Darkness there was: at first concealed in darkness,
this All was undiscriminated chaos.
All that existed then was void and formless;
by the great power of warmth was born that unit.

Thereafter rose desire in the beginning,
Desire the primal seed and germ of spirit.
Sages who searched with their heart's thought
discovered the existent's kinship in the non-existent.

Transversely was their severing line extended:
what was above it then, and what below it?
There were begetters, there were mighty forces,
free action here and energy of yonder.

Who verily knows and who can here declare it,
whence it was born and whence comes this creation?
The gods are later than this world's production.
Who knows, then, whence it first came into being?

He, the first origin of this creation,
whether he formed it all or did not form it,

> Whose eye controls this world in highest heaven,
> he verily knows it, or perhaps he knows it not.

If we interpret the word 'He' as unique and oneness, then it will not be necessary to turn towards an infinite environment for self-cognition.

Hence, a law can be proposed stating that everything originates from infinity and is within Him, and all will return to Him. We cannot consider any name for Him, as there is no name to describe Him the way He is. Existence is at His will, not in a manner that can extend beyond Him in order to imprint creation. This is because for Him no external world exists such that something may separate from Him and manifest beside Him. It is for this reason that He must only be called 'He'.

This matter has been addressed in the Holy *Quran* (2:255) as follows:

> Allah! there is no god but He,- the living, the self-subsisting, Eternal. No slumber can seize Him nor sleep. His are all things in the heavens and on earth. Who is there can intercede in His presence except as He permitteth? He knoweth what (appeareth to his creatures as) before or after or behind them. Nor shall they compass aught of His knowledge except as He willeth. His throne doth extend over the heavens and the earth, and He feeleth no fatigue in guarding and preserving them for He is the most high, the Supreme (in glory).

In any case, since we cannot extract any law from an Unconditional and Infinite Existence, we have no choice but to study the origins and effects of cellular life and other limitations, so that those laws may make us more familiar with life, and make life more familiar to us, even though life is so close to us.

With regard to this matter, I was thinking about the solar system when millions of years ago Mars was in the flow of life, and living in accordance with the laws of balance, capability and resistance, the manifestations of life might have been something similar to what currently exists on Earth. However, today Mars, and all its occurrences and developments in the planetary system, is attached to a deadly silence and is waiting, in accordance with the law of 'returning to the origin', to step towards being archived or towards a new transformation.

Also, the planet Venus is covered with vapour and clouds, and is expanding its orbit so that one day it may possess new manifestations of life and similar developments, like Earth, and pay attention to the book of life and creation. Furthermore, even though it is unaware of its past and future, it recognises effect in place of the cause and does not refer to the cause because of its ambiguity and incomprehensibility and, based on sensory measures, assumes that the existential actuality of a particle is all of Existence.

If a human being sets foot on the planet Venus and discovers the possibility of living there one day, and his innovations make him forget his relics, will he one day presume that there is a possibility that life on Venus may have been brought from another planet? Since it is not impossible to confirm this thought, should we still be certain that single-celled organisms from the bottom of the seas evolved to flat-nosed apes which later transformed into human beings? Or should we be uncertain about every principle we believe in today?

For a correct and wise judgment, and to employ a correct law, is it not better for us to break the boundaries of limitations to attain elevated and truthful cognition, and create a definition for the particle which will be the same as that for the universe?

I have often wondered whether a particle is greater or the universe, and have eventually come to this conclusion that the particle has made up the universe and the universe is of particles and factors which are dependent on and connected to each other.

(Extremely ancient religions like Buddhism, Zoroastrianism and the religion of ancient Egypt claimed that no-one could cognise the Absolute Origin and discover its aims. However, in that immortal part of the human being's existence, there is a reality from the Absolute Origin. It is for this reason that the secrets of the universe should be heard from the mouth of a complete and perfect human being).

Perhaps the ancient sages wanted to say that every being in

Existence is a shadow of its luminous reality. Of course, this is not Plato's Theory of Forms (or Ideas). What I mean is that, when rays of light emanate from a source onto a mirror, other intersecting points are revealed from its reflecting rays. This means that every being possesses a guiding double, with whom he is constantly connected. (What harm is there if it is assumed that the fate of beings has been determined in the heavens? And what is wrong with looking for the Father of Jesus in the heavens*). The human being possesses a luminous reality, whose cognition has resulted in establishing the school of religious cognition. Furthermore, the teachings of most prophets, from proclaiming laws or receiving illuminations, are on this basis.

In the language of scholars of religion and Sufism there are words such as:

- *Nasut* (the material world)
- *Malakut* (the Heavenly realm; the realm of spiritual cognition)
- *Jabarut* (the Sphere of Divine Power)
- *Lahut* (the Luminous Kingdom) within the realm of the Absolute, which is when the identity of the human being is annihilated (which it is) in the Infinite, like a drop of water that disappears in the sea.

(These subjects will be explained, from the current practical aspects covered in the School of Islamic Sufism in a separate treatise).

---

* Heaven in this context refers to the realm of Absolute, which is something opposite to inanimate.

In the works of the ancient sages, it has been said that since God is everywhere, He also is within the human being. We must first try to cognise the human being and then listen to the real voice of God within so that we may be able to hear the commandments of the Absolute Origin.

It is certainly nothing more than scientific justice that scholars of the distant past, and ultimately the prophets, have described the human being in such a way that, with a little attention, his reality must be termed God.

It follows that with this physical body and mind, and our current limited intelligence, the human being will never be able to cognise the Absolute Origin and know how He created the universe and if there is any difference between him and the Infinite Existence. The human being's perception for cognising the Absolute Origin is nothing more than human thought. If we say that one day the human being will be able to describe God, then on that day we will no longer call him human. On the contrary, he will be omnipresent like God. Therefore, on that day we should contemplate that if the human being in its current existence and limitations is the Absolute Origin, or the centre of infinity, and cognises God as He is, which means cognising himself, will it be permissible to say that he has created the universe, or knows how he has created the universe? For in that case we have not praised him as the pinnacle of the essence of knowledge; instead we have counted imperfection, inability and ignorance as his attributes because the current limited human being is in fact restricted to the possibilities of time and space and is himself part of events; and this is clearly different from the reality of 'I' that was discussed earlier.

Therefore, it is obvious that the knowledge of how events happen has no face value before those events occur; or, if the events provide him with any knowledge, how they occur is not within his ability or understanding. Moreover, because the need for knowledge is the outcome of inability, for what and for whom does he want to know? Is there anything besides himself that he wants to know about?

If the complete and perfect human being, whose presence is not superfluous in the context of Existence, does not hear his own voice from within himself and does not recognise it, then what other way is there for self-cognition? These are all issues which have given rise to the cognition of religion and perhaps science, and have found advocates who are amongst the deepest thinkers of the world.

A question arises here: "What should a human being do in order to be as Existence intends?"

The answer to this question is: "Should he do anything extra?" and "Do goodness and righteousness mean anything other than the reality which he has?"

Are these explanations not the final answer to the trials set by humanity for such thoughts? Do we not have a right to announce to humanity that it should look within in order to discover the mysteries of the universe, and not to forget any angle of his inner and outer being whilst conducting research so that he befittingly cognises reality; and to feel the effusions of the Absolute Origin and the Supreme Almighty God with his conscience?

At this point it is necessary for me to add short notes on scientific divisions and the undoubted numerous spiritual veils, which are within the category of the teachings of my Master, my father, and are presented here mostly in the form of excerpts from his book *Az Janin Ta Janan* (From Foetus to Paradise). My Master has precise classifications regarding the set of bodily and spiritual organs of the human being, which he describes:

> The reality of the human being can only be discovered with the 'heart's eye'. Ultimately, that which forms and supplies the existence of the human being is neither material essence, nor protoplasm, nor the cell and nor these strange, amazing, abundant compositions of carbon, hydrogen, oxygen and sulphur; rather, it is solely that non-material, invisible life-giving power. It is this which has created an assembly of an uncountable number of particles and has brought about the astonishing organisation of living matter, which it is directing. Matter and energy have never been separated from each other; the existence of each one involves the other.

Decomposition of the body after death is in reality a continuation of that gradual endeavour which was in progress during life when worn out particles were continuously being replaced with new ones. The immortal and organising spirit is the centre of this power. What we see on the surface deceives us. The truth is that which cannot be seen.

Besides particles with certain chemical and physical properties there is another non-material force within the existence of the human being which is an independent entity called the spirit. However with respect to this, physiologists say that in the same way that gravity, heat or electricity would not have an external existence without a heavy, hot or conductive body respectively, life, sensitivity and thought could not exist without a living and attentive human body.

Therefore, the spiritual works manifested by human beings are born from the neural tissues of the brain. We previously stated that matter and energy necessitate one another in order to become the source of any manifestation. In Existence, however, each one has an independent definition. In any case, with regard to the illogical reasoning of Materialists, it is better to understand matter first.

According to public opinion, matter is something which can be felt and weighed. Yet we observe something in humans which is neither tangible, nor weighable, nor visible; and this is a spiritual entity, which thinks and makes judgements between right and wrong. When awake, as well as when it is asleep, it sees and hears from far distances and imposes its will on others. Moreover, it gives accurate information regarding its own future and that of others, which no one, not even himself, has knowledge of.

More astonishingly, it gives information about events which occur some years later. The map of these events has not yet been planned in our world for it to be reflected onto the brain of an observer, and consequently run onto his thought and later his tongue. In all these inevitable instances, the tangible materials of the body have neither interfered nor helped; and the cells could not have been able to foresee the future events.

If obduracy and prejudicial ignorance do not darken or blind our true sight, we will be awakened by witnessing these unexpected events and we will realise that the principle factor and the real commander of all human thoughts and deeds is the spirit and its spiritual cognitive power. The nerves and organs of the brain are no more than tools; we are the manual facilitators and the Truth is the active power.

If, as in the opinion of the Materialists, the inception and discharge of thoughts from the brain is the same as the secretion of bile from the liver, then it should follow that in old age and at the time of fatigue or brain surgery, these powers should be destroyed or at least deteriorate. However, many scientists become more experienced and mature with age, and their thoughts are more beneficial in old age than at other times in nurturing the younger generations.

As I am writing these words, I realise that it is my spirit which thinks and compiles appropriate words and eloquent expressions to convey meanings and examine subjects. To a large extent, it also ensures that my prose flows freely and articulately. It is not the nerves and muscles of my body, nor the fingers of my hands or other similar cells. This is because I am the imperious owner and commander of this body; the body does not have authority over me.

An Essence, which has not found a part in Existence
Will it ever be able to bestow Existence?

Have you not seen that many patients and illnesses showing resistance and tolerance towards medicines, and remaining unaffected by them are, in contrast, affected by and cured through spiritual healing, human will, concentration or affective prayer and submission?

- The miracles of the prophets
- Magnanimities of the Saints
- Revelations and similar spiritual wonders or innate mysteries, in whose manifestations no outward instruments are involved
- Sacrificing one's own self to save others, when according to the law of nature loving one's own self is inherent within each of us
- Bravery and not taking seriously adversities and the fire which burns the flesh
- Pity or cruel-heartedness

- Love of the Saints
- Freely going to the battlefield, which is generally against natural desires and contrary to the material properties and structure of the body

These are all powerful proof of the free will and independent identity of the spirit and its incongruity with material activities, including the chemical, physical and natural properties of the brain.

The beating of my pulse and the rhythm of my breath, even in my sleep or unconsciousness, where my will is not involved, are proceeding continuously with their duties. However, in talking or writing, my tongue, hand and pen will not move until I will them to do so. I must use my brain to think, make my hands and tongue move, pay attention to the good or bad nature of the topics and judge the credibility of the words in order to extract a result and put it into action. All these activities stop as soon as I change my mind and avoid that topic.

If these works were born from the neural tissues of the brain, like the heartbeat, pulse, breathing and other organs, they should be ceaselessly at work, regardless of my want or awareness, just as my will has not interfered and will not interfere with the natural activities of the body. Heartbeats are continuous, but the opening and closing of the fingers happens when I want to hold a pen and draw the line which is in my mind. Therefore, the motions of the pulse and fingers have two motives: One is unconscious and the other one is conscious.

If we think impartially for a few moments and ask ourselves: "Have all these powers, arts, astounding and extraordinary inventions of the human being come from these six kilograms of bones, or fifteen kilograms of organic and inorganic compounds and fifty kilograms of water? Is it the substance of the human brain that has modified the limitations and defects of its barbaric era to reach today's civilisation and has manifested these sublime senses?"

The answer is most certainly negative and we may remind ourselves that the development and improvement of anything is through the passing of time, with constant care and attention. However, we saw and realised that all cells in the body, without exception, wear out in a short space of time and are replaced with new cells. Not even one cell has remained from the past work on pushing barbarism towards civilisation.

Nerves and the structure of the eyes are merely acting as a pair of spectacles, which at best can be a telescope. Is it the telescope that perceives the planets' and asteroids' orbital journey, chemical properties, physical activities and their other qualities? Or is it the spirit and thought of the astronomer who uses his eyes together with spectacles, telescopes or whatever else is required in any way he desires?

Of course nerves and muscles, which are the tools and appliances for life, are useful and effective; however, they perform their functions under the supervision of the spirit.

The human body is not dissimilar to a car which, despite the availability of a chassis, an engine, petrol and other parts, is unable to move without a driver. Supposing that for any reason it did start to move in the absence of a driver then its outcome is obvious.

Order, discipline, science and inventions, as well as the provision of medical facilities to guarantee the well-being of the body are not the products of cells and molecules. Intellect, law and power are needed to control such an organised world. Everywhere, the law is a sign of the existence of the legislator; and discipline and order are the reasons for the existence of the organiser and the governing intellect.

If the course of creation were to be other than what we said, i.e. without a guardian, or the car without a driver, the world and whatever exists within could not have been protected from the hazards of accidents. In any case, the material world everywhere, as well as in the human body, is the means by which laws set by an impeccable legislator that wisely controls and organises 'the great machinery' of creation are performed. In reality, the human being is a moving atom positioned on another moving atom (the planet Earth) and the performer of this organisation is a powerful encompassing force, whose manifestation is the human being (the realm of microcosm).

After splitting the nucleus of the atom in this century, and its magical effects, the Materialists have also inevitably moved away from their unacceptable thoughts on worshipping matter and witnessed the governance of energy on matter with equipped eyes. Hence, they have come to believe that there is nothing acting in the universe except the waves of energy, and that the differences in appearance and colour are the result of the number of atoms and their density as well as the scale of their vibrations, whose speed is beyond normal comprehension. The retina of our eyes senses those vibrations and perceives every being in a particular form.

So the saying of the Sufi (*Aref*) who stated in the past that 'Existence is One', has been true and correct just like his other beliefs. Our outward-seeing eyes cannot perceive the real identity of the human being and are ignorant of the power and the faculty of the intellect which are dominant within him. The spiritual faculty works in association with the physical matter of the brain, but its life is other than that of physical matter.

FIVE
THE SPIRIT AND ITS CREATIVE WILL-POWER

Various different religious faiths and philosophies believe in the existence of angels, celestial beings and other similar entities (like the Essences [*Farvahar*]of Zoroaster and Forms [Ideas] of Plato). If it is considered in fairness, the human being has not yet cognised himself and has neither discovered the truth of his essence nor the centre of his will, action and power; and is currently on a journey of gradual evolution.

Unfortunately, the classical sciences only consider and analyse the most superficial aspects of the physical structure of the human being. They do not pay attention to the precise, deep and subtle masterpieces of this mysterious creation, and its great inner organisation and the higher levels of this sublime palace. In every century a limited few have been able to travel the inner world, and they are always in the minority.

Materialists say that even though some of the unknown powers of the brain can be developed by certain exercises, which become the source of extraordinary happenings, it does not mean that there is an independent spiritual state. Yes, with exercises, which strengthen and clarify the five senses, the vibrations and waves from the world outside the brain can be discovered more clearly. However, for these discoveries, the brain requires certain tools, and these essential tools include the senses, whereas in fact the processes of will-power and thought are completely different from this. This means that vibrating motions exit the brain and become dispersed in space. Therefore, there must be another cause for the vibrating motions emanating from the brain and that cause is the capable and independent identity within the spiritual centre of the human being.

The effects of spirits from a far distance, the propagation of thoughts, the discovery of a star's structure by analysing its light and spiritual manifestations at the time of death are similar to the magnetic power of iron, albeit the spiritual connection of the human being is a more exact and superior form of this. These truths will be properly discovered when this precious phrase of 'cognise yourself', which existed as the objective of sublime thoughts even before the Delphi temple and until now is commonly agreed upon amongst Eastern and Western scholars, becomes completely practical and yields results.

Each one of these truths demonstrates that the spirit and the existence of the human being are not born out of matter. On the contrary, matter is a product of the soul which it has created to aid in fulfilling its purposes and as its tool. It is not the optic nerve and the brain which observe events before they occur externally, as this is contrary to the laws of nature. Not only do they see far and from behind the veil in the state of somnambulism, but some people also observe this when fully awake. More amazing still, sometimes they even 'see' with the forehead and the fingertips ('The Sufi's entirety becomes all seeing').

With all these amazing phenomena, is it appropriate to suggest that the grey matter of the limited brain - whose cells are continuously wearing out and facing annihilation and are being replaced with new cells* - is the generator and the origin of this exceptional power, and the encompassing force within the whole body, and to consider it as the effective agent in the greatness of this peerless organisation?

Seeing future events is not a physiological quality. It is a faculty higher than nature. This means that seeing future events is a 'vision' which has occurred in the spirit but has not yet been revealed onto the material world; and as we say, 'has not been created', and has not put on an 'earthly cloak' and, on the path of creation, it has not yet reached the physical matter of the brain for it to start working and surge out of the brain. How can the limited brain be able to inform of events which do not exist? Therefore, it relates to a power higher than nature.

---

* Hazrat Shah Maghsoud Sadegh Angha confirmed the process of brain cell regeneration, or neurogenesis, 26 years before it was scientifically proven in the laboratory. At the time when *The Hidden Angles of Life* was written in 1972 (first published in 1974) scientists widely believed that neurons were not added to adult mammalian brains and as such the theory of neurogenesis was mainly rejected. Eventually in 1998 neurogenesis was confirmed in the adult human hippocampus. PS Eriksson, E Perfilieva, T Björk-Eriksson, AM Alborn, C Nordborg, DA Peterson, FH Gage, "Neurogenesis in the adult human hippocampus". *Nature Medicine*. **4**, 1313-1317 (1998).

SIX
CLAIRVOYANCE

Any event which must come into existence has a certain appearance in the preceding realms, according to the spirituality of that place; the journey of the soul of man is also based on this foundation. For example, we commence our journey from the realm of the Luminous Kingdom (*Lahut*), Archetypes (*A'yan*) and 'the Guarded Tablet' (*Lawh-i-Mahfuz*) down to the realm of the material world (*Nasut*), which is the lowest stage and where we put on our sensory clothes. That spiritual event has also accompanied human beings through this course, riding the magnetic waves until reaching the material brain where it is manifested. Therefore, it can be said that a clairvoyant person sees an event with his Heavenly eye (*Malakut*), with which he has been travelling on the same course, until its turn arrives to be witnessed by the material eyes and brain. After it has reached the world of elements and matter, the human eye observes that event again. The gnostics, who do not neglect their Heavenly realm (*Malakut*), have always encountered such observations and are familiar with them. It is for this reason that they sometimes foretell events that have not yet occurred.

Each one of the great gnostics has benefited from and received a portion from these Heavenly gifts according to the vastness and purity of their hearts, and their clear and skilled minds and will continue to do so. They deduce and comprehend foreseen affairs, whose knowhow is necessary and noteworthy for them, via the path of instinct and revelation; without the need for logical rationalisation, relying on conventional science or resorting to common apparatuses and analyses.

The great philosophers of the world have spent much time and attention on the issues surrounding spiritual matters but have been unsuccessful in obtaining definite answers. Conversely, those who possess illuminated hearts are completely successful in discovering the secrets of topics with their revelation receiving minds via the path of enlightenment. This is because this group of scientists of spiritual cognition and Illuminationism are drawn towards the path which leads to new secrets. Clairvoyants are capable of reading the thoughts of others without using any of their natural sensory organs. Moreover, they see future and far distant events even before they occur (as was explained earlier) prior to their external manifestation.

Clairvoyance is similar to revelations, religious states and telepathic qualities.

Unbounded beauty and love are invisible in the heart of the visible world, and only deal with pious people. Moreover, this state of piety is essential for the ascent of the spirit to the source of Existence and the centre of Divine Power and Will. It will not result without discipline and constant strife with the carnal soul and overcoming worldly attachments. For this reason, seekers of the truth are few and advocates of obstinacy are many. That aspiring human being who undertakes this great matter breaks all the chains of attachments and liberates himself from all worldly attractions; and in those dark, blurry nights, as he sits regretting the past, weeping for his weaknesses and errors, with presence of heart he asks for forgiveness from the source of God's mercy. He bears the hardships encountered during purification with delight and excitement until gradually with the help of the pure spirits he unties the knots tied by satan and reaches the state of real liberation. Moreover, his supplications take on the form of a blissful journey until they are accepted.

The spirit of a Sufi is outside the limits of time and space. It is connected to those connotations and mysteries which go beyond our limited vocabulary. It culminates in unification and monotheism and in reality stays with his God.

These gnostics see with the eye of the heart what others cannot see and know what others do not know. They possess a creative power of which the materialists and superficialists are unaware.

We must not have bad intent towards these gnostics and their miracles and should satisfy ourselves with an experimental concept.

We must realise that, just like Divine revelations, this blessing is the very truth.

SEVEN

VARIOUS RELIGIONS AND ANALYSIS OF THE INNER POWERS

Islamic Sufis consider the heart to consist of seven "Mounts" (*Atwar Sab'a*):*

- The First Mount is the Chest (*Sadr*):

  Is one whose heart Allah has opened to Islam...

  Holy *Quran* (39:22)

- The Second Mount is Luminosity (*Lam'an*), the light of intellect and faith:

  ...he has written faith in their hearts...

  Holy *Quran* (58:22)

---

* In a collection of four treaties presented on the essence of Islamic teachings and Sufi practices entitled *Al-Rasael*, Hazrat Shahmaghsoud Sadegh Angha states: "the Enlightened of the Abode of Eternity have described the states of the heart in respect to the seven heavenly spheres thus: its stars as the light of the stages, and its ascendance as the spiritual ascent of the journey of the salek; the seven mounts of the heart correspond to the hidden lights of the sky of the heart that by divine inspiration and absorption in the rapture of love, the heart is illumined and freed on its flight to its Origin". (Hazrat Shah Maghsoud Sadegh Angha, *Al-Rasael*, M.T.O. Shahmaghsoudi® Publications, 1987).

- The Third Mount is the Membrane of the heart (*Shaghaf*), the place for the light of affection and love:

  ... indeed he has smitten her to the heart with love...

  Holy *Quran* (12:30)

- The Fourth Mount is the Heart (*Fowad*), the place of seeing:

  The heart lied not (in seeing) what it saw.

  Holy *Quran* (53:11)

- The Fifth Mount is Love from the heart (*Hobbat al-Ghalb*), it is the place for the manifestation of the love of God:

  ....they love them as they should love Allah, but those of faith are overflowing in their love for Allah....

  Holy *Quran* (2:165)

- The Sixth Mount is the Core of the heart (*Soweyda*), the source of wisdom and light of Divine knowledge.

- The Seventh Mount is the Secret (*Serr*), the place for the manifestation of Divine lights and secrets.

In the book Hamaseh Hayat (The Epic of Life), in addition to these seven stages, I have also written about seven lights, and have mentioned all the necessary spiritual aspects therein.

Leibniz, the German philosopher, labelled 'essence' as a monad, and has said that monads are metaphysical. Their active powers consist of 'Perceptions'* and their differences result from different 'Perceptions'. He says that God is the eternal monad and other monads are illuminations of His light; just as the Essence (*Farvahar*) of Ahura Mazda is the greatest among the essences.

The body which we see is a collection of many monads. The Essence (*Farvahar*) of Avesta, the Monads of Leibniz and the Forms (Ideas) of Plato are the principle moulds, sources of meanings and the true elements of all beings. They control matters, and provide the power of life for creatures. Moreover, for every single entity in the world of Forms (Ideas) there is an Intellectual being, which acts as a guiding teacher and god, providing power and protection.

The rational soul of human beings comprehensively comprises all the senses and the spiritual cognitive faculties, and in fact instructs and governs bodily affairs, as well as the affairs of the vegetal and animalistic souls which are functioning within the human body. The human being is the epitome of all these powers and manifestations. Moreover, the rational soul of man is a god for the lesser souls where each one of them governs the body and its organs, much like the provinces of a country. Each one of the senses, the spiritual powers and the inner powers of man are also by their own right, an independent ruler in

---

* In section 14 of his *Monadology*, Leibniz explains that "Perception", which must be distinguished from Apperception or Consciousness, is a momentary state involving and representing a multiplicity in the simple substance (i.e. the monad). (GW Leibniz, *Leibniz's Monadology / translations by Nicholas Rescher*, University of Pittsburgh Press 1991)

their own domain. The human soul is the monarch, the unconditional ruler and the greatest potentate over the smaller governors. In spite of all these independent and different realms, and numerous centres, the rational soul still remains in its oneness, and all this multiplicity does not disturb its unity.

That which is meant by:

- The Essence (*Farvahar*) of human beings according to Zoroaster
- Platonic Forms (Ideas) of Platonists
- Active Intellect in The Philosophy of Illumination
- Ancient Adam in Jewish Mysticism
- The Perfect Human Being in Islamic Sufism
- The Angel Gabriel and the Spirit in Islam
- The Holy Spirit in Christianity
- Atman and Purusha in Theosophy and Brahman Philosophy

is an Absolute Essence which is a Divine educator and guide for the rational soul of man. In modern sciences, it has been said that the innumerable types of matter are produced from simple chemical elements according to the number, composition and movement of atoms. Moreover, in limbed-animals, the cells also create many kinds of compositions and forms of organs according to a special order.

The state for all beings in all realms accords to this convention.

This means that the faculties, essences and lights from the invisible world, like the cells of organs and atoms as an example, form groups, clans, tribes and nations, the smallest of which start from two faculties, or two essences, until they reach unimaginable numbers. Each one of the faculties in the three realms of the material world (*Nasut*), the Heavenly realm (*Malakut*) and the Sphere of Divine Power (*Jabarut*), or in the region of Nature, Soul and Intellect, are guides and driving forces for the subordinate world.

Furthermore, there is a faculty beyond each of them until they reach the Will of God, which is itself the driving force and guide for all stages within the realms of Existence. That Omnipotent has placed a burning love within the heart of every being, for its elevation to higher ranks. The substantial motion and perpetual attempts of the particles towards higher realms is the effect of this very love, which is the motive and axis for the vicissitudes of life's cycles and for the power of living.

It is this very attraction of love which has, both outwardly and inwardly, connected and joined together all the beings, especially the human being, within the realms of Existence. Furthermore, it has created a perfect unity within multiplicity which is so organised that if one part of or one group of particles does not act in harmony, it will culminate in distress, agitation and lead to destruction.

The spiritual superiority and the existential value of each being is proportional to the intensity of its love, its sacrifice and the merits of its voluntary actions, which is connected with the effects of cognition. The supreme manifestation of this superiority is the human being (*Adam*) whose body represents the whole of nature, whose soul represents the Universal Soul, and whose intellect represents the Universal Intellect; and who is the mirror of Divine deeds, virtues and essence. In other beings, these three faculties are still in a state of waiting. Much time, great endeavour and extensive development are required before they can completely leave the bounds of inanimate objects, plants, animals and beasts in order to possess '*Adam*-like' virtues.

As medical science has demonstrated, the human embryo in the womb passes through the inanimate, plant and animal stages with exceptional speed.* In addition, many of the natural forces and powers such as magnetism, electricity, attraction, repulsion, memory and so forth exist, and are working within the human body. Even the phenomena of thunder, lightning, storm, water, wind, fire, and whatever is in the world of nature, exist in different forms in the human. They are all performing their duty in all the particles of the body, individually and collectively, according to the principle of order

* Reference to Ernst Haekel's "biogenetic law" formulated in 1872 stating that "ontogeny recapitulates phylogeny", in other words each of the successive stages in the development of an individual represents some aspect from its evolutionary past. For example, although all vertebrate embryos pass through a stage of developing embryonic "gill slits" they only become true gills in fish; while in other vertebrates they become part of the jaw or ear apparatus. The "biogenetic law" had been neglected for much of the 20th century. A recent surge of interest in this area has given rise to a new field of research, which combines developmental biology with evolutionary concepts, often referred to as EvoDevo. In a recent study scientists used a computational system, exhibiting evolutionary dynamics, to demonstrate that the "biogenetic law" holds (for digital organisms at least) since "traits that arose earlier in a lineage's history also tended to be expressed earlier in the development of individuals". J Clune, RT Pennock, C Ofria, RE Lenski, "Ontogeny tends to recapitulate phylogeny in digital organisms". The American Naturalist **180**, E54-E63 (2012).

(like our solar system), based on cooperation and harmony. The human being, or other manifestations of creation, are all inundated with the attraction of love and will not abandon the essence of love.

Therefore, in the entire world of creation there is one intellectual power and one spark of love hidden within the essence of every particle, which is its driving force.

There are four types of 'tools' available in this world for the rational souls to gain knowledge of the truth.

The first is the material organs of the body for passing through the material world (*Nasut*). The second is the five senses (some recent opinions believe in ten* or more senses) for passing through the lower vicinity of the Heavenly realm (*Malakut*) (which is within the proximity of the Earth). The third is the inner faculties, or intellectual faculties, for passing through the higher environment of the Heavenly realm (*Malakut*) and the outskirts of the Sphere of Divine Power (*Jabarut*). The fourth is spiritual faculties and Divine power, specific to the completely sacred souls, for passing through the Sphere of Divine Power (*Jabarut*) and the Luminous Kingdom (*Lahut*).

The material world (*Nasut*), which is the lowest stage of the

---

* Neuroscientists still debate as to what constitutes a "sense", and as such some argue that the "sense" of balance (equilibrioception), heat (thermoception), pain (nocioception), body awareness (proprioception) and the sense of time should be included with the five standard human senses (vision, audition, gestation, olfaction and tactition).

hierarchy, is located in the middle (centre) of three higher and three lower worlds (inanimate objects, plants and animals). From the material world (*Nasut*), in an ascending arc, the rational soul traverses the three higher levels, one after another, and returns to its origin. This relates to the saying of the Holy Prophet of Islam that everything returns to its origin.

It is not possible for the delicate rational soul to mix with the harsh terrestrial and physical body as a means of travelling and perceiving the realms of spiritual cognition. Just as travelling by air or water in the material world (*Nasut*) requires the suitable tools to be lighter than air or water for a safe journey, apart from the physical body, which travels in the earthly world, the rational soul also requires three more subtle bodies – as described below - with which to travel through the three worlds [Figure 3, page 152].

1. The etheric, or corporeal, body which is specific to the sensory realm of the material world (*Nasut*).
2. The astral, or celestial, body which is specific to the realm of Reality and the Heavenly realm (*Malakut*).
3. The rational, or causal body, which is specific to the contemplative sciences and the Sphere of Divine Power (*Jabarut*).
4. The luminous body, or the fire halo, which is specific to the realm of the Luminous Kingdom (*Lahut*) or the manifestation of Essence.

1. The Etheric (or Corporeal) body

This is the first vessel of the rational soul and is formed from etheric elements prior to the formation of the embryo . It is created by the hands of angels (the power of creation) and supreme beings, as a template from which the physical figure of the body is shaped. This etheric body which surrounds our body like a thin veil and a cloud (or phantom) - similar to a thought - is sometimes seen by clairvoyants. This halo does not detach from the perimeter of the body, except on special rare occasions.

The role of the etheric body is to receive the power of life from the sun and to deliver it to the body (this was previously discussed in the section on the magnetic powers and its cosmic connection). The health of the physical body depends on the health and excellence of the etheric body; and therefore, since this is made of etheric elements which are considered to be more subtle than air and are the subtle parts pertaining to matter (the principle matter of life), in spiritual summoning the audience's etheric matter, particularly that of the medium who is more prepared, is used to manifest flowers, fruits and in showing parts of the figure of the spirit (for example its hand, feet and face). For this reason, it has been said that the presence of angry and weak individuals during such spiritual summoning is harmful - especially if the mediator is ignorant of these subtle points or is psychologically ill. The etheric body, which is responsible for obtaining the power of life (*Prana*) from the sun and transmitting it to the body, will not remain for more than two or three days after the spirit has separated from the body (perhaps it is the etheric body which perishes shortly after death). It is by means of this etheric body that objects possess attractive and magnetic properties (living beings have this attraction).

What is observed around graves after burial of the dead, similar to ghosts (phantoms), is this very etheric body. This etheric body, which is also called the etheric halo, visibly extends by approximately four finger widths out from the body. In natural states, it can sometimes be seen to be extending out from the skin of the body by several finger widths. When the physical body is washed, the etheric body is also cleansed.

The etheric body is composed of the four elements, and the quantity of each element differs according to race, tribe and nation.

Clairvoyants observe the colour of the etheric body as grey-violet or dark grey. Its pale and dim colour, as well as its visibility, is relative to the subtlety or harshness of the body. The etheric body is also the intermediary between the astral body and the physical body. It conveys the perceptions of the body's sensory relations to the astral body; and it also descends and distributes the perceptive power of the astral body, and stages superior to that, to the brain and other parts of the neural system. Separation of this body from the physical body is very difficult for normal individuals. However, this is very easy for individuals who become intermediaries (mediums, or enchanted people) between the Celestial or Spiritual realm.

Numbing an organ, which is customary in surgery, expels the etheric body from that organ; and consequently the connection between the perceptive faculty and the organ is cut as a result of the numbing.

The relationship between the etheric and physical body is so exact that, if an injury was sustained by the etheric body, its effect would manifest at that same point on the physical body. This effect would also appear in the astral body. Although the etheric body cannot be seen with ordinary eyes, it is of material substances. Therefore, warmth, cold and tastes affect it. It has been weighed on sensitive scales, and it is for this reason that the body is lighter after death than before it. This means that as a result of death a material substance, which has a mass, leaves the body.

2. The Astral (or Celestial) body*

The rational souls of most of humankind can only travel and ascend through the Celestial realm or the lower vicinity of the Heavenly realm (*Malakut*) - which is connected to the material world (*Nasut*) - with the astral body. Therefore, the circumstances surrounding this body are worthy of more attention.

In general, the rational souls have brought the four mentioned bodies with them from the realm of particles to this world, because it would not be possible to live in this world without them. The difference between the four bodies relates to their luminosity, subtlety, weakness and the power related to the degree of the soul's evolution. By perfecting their power and subtlety, they will ascend to the sublime stages. Just as the etheric body is nourished and nurtured by materials from the material world (*Nasut*), the astral (or celestial) body, which is the offspring of the Celestial realm and the Heavenly realm (*Malakut*), is also nourished and nurtured by celestial conceptions and a range of spiritual cognitions.

Elements from the Sphere of Divine Power (*Jabarut*), meaning Absolute Comprehensions, also strengthen and perfect the causal body. This, too, is the case for the luminous body, whose nourishment comes from the realm of the Luminous Kingdom (*Lahut*) through the vast sublime truths and the bounties from the essence of Divinity.

* Other names for this body include "the mist of prime matter" (*hoor-e ghalyaie*), or "the heavenly mist" (*hoor-e Malakuti*).

The physical body is heavier, more material and darker than all the other bodies. It is located in the middle (of two descending and ascending arcs). The etheric/corporeal body which is slightly lighter is above it. The astral body which is even more subtle is above the etheric body, and the causal body is above the astral body. The luminous body, which is more luminous and subtle than all the others, and is the furthest away from matter, is situated above all and encompasses all the realms.

Natural born clairvoyants, or those who become clairvoyant or clairaudient through austerity, see the halos of the four bodies around the physical body coloured in proportion to their subtlety and luminosity. They can even distinguish sudden changes in colours and luminosity, which are produced from the waves of emotions and admirable and disagreeable feelings. Those who possess admirable feelings attain a shining astral body which prepares them for ascension towards sublime stages and familiarity with the spiritual and luminous beings. They attain many mysteries and secrets in this way, and encounter the Heaven of spiritual cognition. Conversely, hellish attributes will face towards hell.

The three luminous realms are like three circles inside one another which, with cooperation, are heading towards one goal. The situation is the same for travelling through the Celestial, Rational and Luminous realms, which are in complete unity. Each one has an effect on the other according to its power. One of the 'clothes' for travelling in the realms referred to earlier is the same as that which is used by the rational soul during sleep, at the moment of the soul's temporary release from the body for travelling through the Heavenly realm (*Malakut*) or the Sphere of Divine Power (*Jabarut*).

It is with this very astral, or celestial, body that manifestations occur in spiritual summoning circles of every country, and which has been confirmed by famous naturalists. Since this astral body is the centre for all the pure and evil feelings and qualities of human beings, it will be associated with his body after death and will enter its own firmament until it discovers the outcome of his own pure and evil deeds. Therefore, any recompense or requital will be related to the subtle bodies of the rational soul.

Clairvoyants say that due to its subtlety, the astral body constantly remains in their view as moving, undulating and luminous, similar to steam or a bright cloud. It is for this reason that it has been named the 'astral body' in European languages. This body does not possess parts. Rather, it is like a cloud that can be seen from a far distance. Therefore, it has no foundation or activity in the majority of humankind. The upper part of this body, which extends from the crown of the head to below the chest without a lower part, is often seen by clairvoyants during wakefulness as appearing similar to phosphorous and light.

3. The Causal (or Rational) body

The third subtle body with which the rational soul travels to the Sphere of Divine Power (*Jabarut*), Comprehensions and all Truths is the causal body. This body is essentially more luminous, translucent and subtle than the astral body; and it is an indication of the measure of the human being's evolution of thought and intellect.

The sinew of the causal body is woven from the perceptions, thoughts and wishes of the rational soul. Moreover, its luminosity and subtlety is proportionate to the purity of intentions, ideals and moral virtues.

The aforementioned bodies possess seven degrees and each degree has a magnanimous and nurturing position in relation to its inferior. Thoughts and Comprehensions in the higher level of the Sphere of Divine Power (*Jabarut*) are extremely luminous, vast and Absolute. The more they descend the more they lose their vastness and absoluteness until the seventh degree, which is tangential to the lower vicinity of the Heavenly realm (*Malakut*) and where they take on the visage of the mind. Also, the closer these infinitely fast waves get to the lower degrees, the more sluggish and weaker they become; until they reach the Celestial world and eventually reach the final degree of callosity and condensation in the world of nature, where they transform and become tangible.

It is with this causal body that perfect souls are able, with creative will and concentration of thoughts, to keep the visage of the mind and intellect constant for it then to take on an exterior and material existence. Revelation and Inspiration are also related to this very Rational and Absolute realm.

The descriptions of the qualities of the Sphere of Divine Power (*Jabarut*) and the causal body have also been highlighted in the teachings of Theosophy. The spirits of the Sufis, Saints and those possessing moral values, as well as the disciples of the path of spiritual cognition, perform their duties and obligations in guiding souls on this earth and directing the spirits who have recently entered this realm proportionately to the perfection they yield in the upper and lower levels of this realm. The real abode of these guides and purified teachers (invisible dignitaries) is the Sphere of Divine Power (*Jabarut*). From here they inspire and distribute the lights of blessings and luminous thoughts, and spiritual cognition and Divine ideas, to the core of those in search of virtue and truth and to the students of the path of spiritual cognition, science, philosophy and fine arts.

The Sphere of Divine Power (*Jabarut*) is divided into two parts: a higher part, consisting of three higher degrees of this realm and a lower part consisting of four lower degrees. That subtle body, which has the right to reside in the higher part of the Sphere of Divine Power (*Jabarut*) is called the Constant body, or the causal body, which is the sum and repository for all durable experiences of the rational soul for future living. For this reason, it has been called the causal body (i.e. the centre for all causes and origin of life of the rational soul). The causal body is the vessel and constant 'clothing' of the rational soul, which it never wants to, nor can, remove. For the spirits, this body is like the skin of the physical body.

4. The Luminous body or the Fire halo

The ascent of the rational soul to the realm of the Luminous Kingdom (*Lahut*) is done with this body. Since with great effort, much patience and gradual evolution the soul of man can reach this rank, it will not be possible for any lackadaisical, impatient and desirous person (it requires fervent love, which is the astrolabe of God's mysteries). The more that the rational souls devote to serving and guiding other beings, and the more in love they are with reaching their own purpose, then a veil of light will be formed, to that same extent, around the causal body, like the halo around the moon, which is called the fire halo (the luminous halo which is portrayed by painters around the heads of prophets and the pious is based on this connection. In Buddhist scriptures it has been written that the halo around Gautama Buddha's head was so vast and bright that it could be seen from several kilometres away). The reason for painting this luminosity around the head is that the centre for the manifestation of the Celestial lights of the Intellect of the Sphere of Divine Power (*Jabarut*) is located in the infinitely small glands of the brain, in the spinal cord and in the pituitary gland.

■■■

I think I have sufficiently explained the propounded subjects. In case any difficulty arises during reading the topics of spiritual cognition, which is discussed in the School of Islamic Sufism, they will be analysed and explained in a paper which I shall write as a compliment to this article. It is necessary to bear in mind that this submitted article is based on questions set by a scientist like you; and perhaps in the future I may be able to address more openly my personal discoveries gained from researching life and even elucidate the real reason for the creation of the School of Islamic Sufism and how I am spiritually connected to it, and how I do not know any other way for civilisation except through the development of the School of Islamic Sufism in human societies. Please accept my sincere appreciation of your commitment to scientific research.

Shah Maghsoud Sadegh Angha

'Oveyssi'

December 1974

Tehran - Iran

Figures

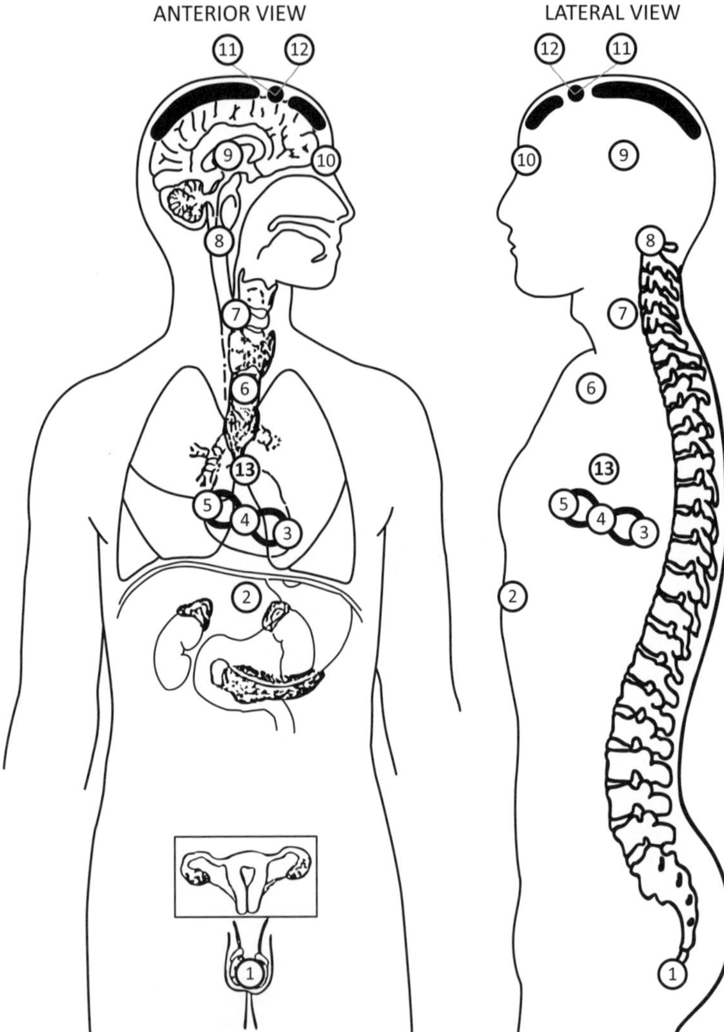

1. Coccyx	2. Solar plexus	3. 4. 5. Heart
6. Thymus	7. Throat	8. Brain stem
9. Third ventricle	10. Third eye	11. 12. Crown (ant.fontanel & grey layer)
13. Source of Life		

Figure 1: The human electromagnetic centres and the Source of Life

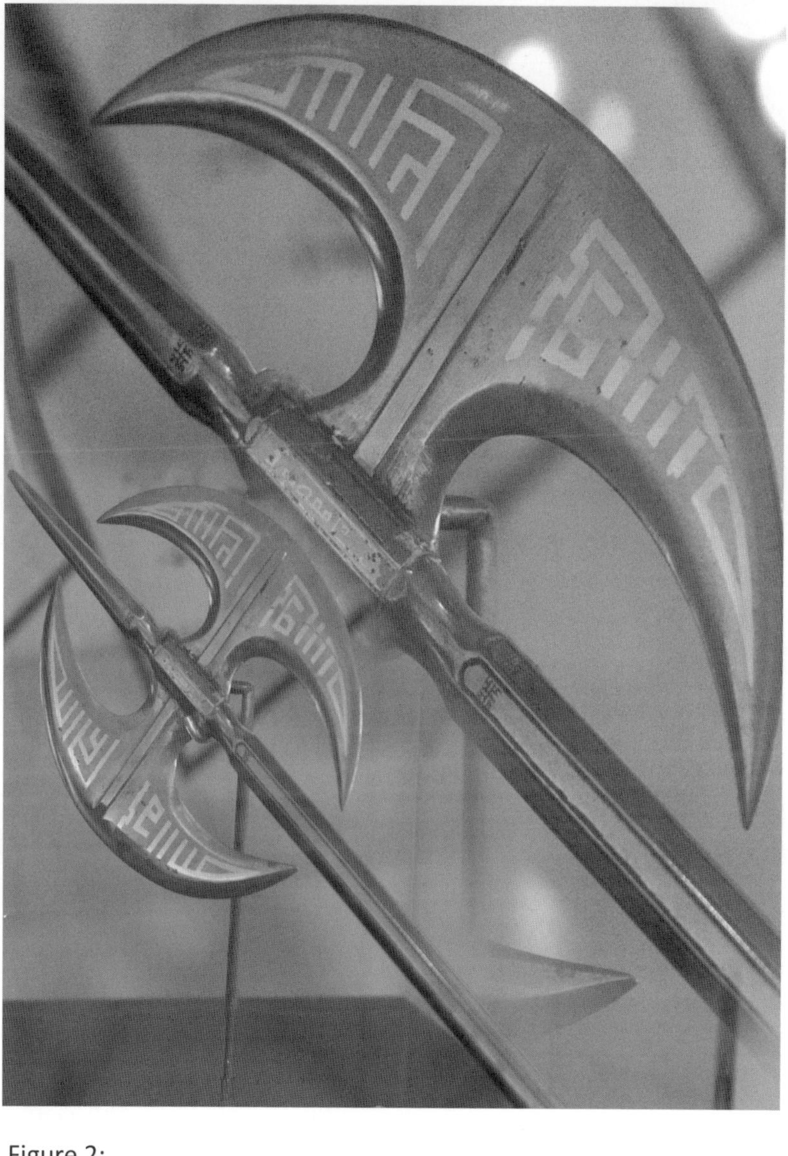

Figure 2:
Photograph of the actual *Tabarzin* referred to by Hazrat Shah Maghsoud Sadegh Angha in the text; currently on display at the MTO Shahmaghsoudi centre in London, UK.

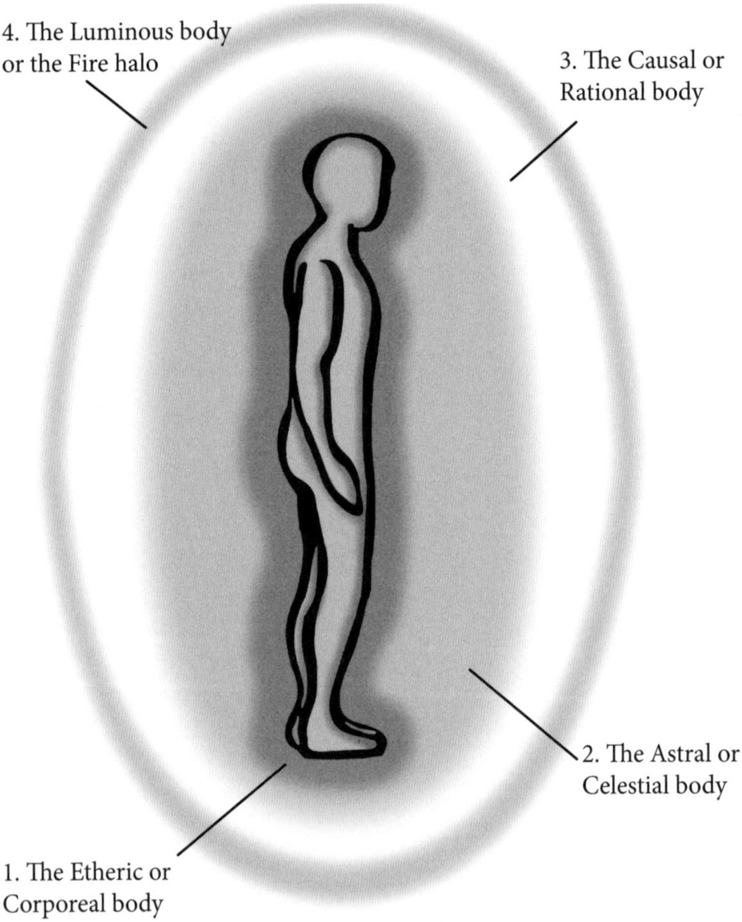

Figure 3: The four subtle bodies of the rational soul

Appendix

Appendix I

Verses from the Holy *Quran*

Surah	Chapter	Verse	Page Number
Al-Baqara	2	255	106
Al-Baqara	2	165	130
Yusuf	12	30	130
Al-Hijr	15	16-30	102
Al-Hajj	22	46	63
Az-Zumar	39	22	129
Ash-Shura	42	24	64
Qaf	50	37	76
An-Najm	53	11	130
Al-Mujadila	58	22	129

Appendix II

Verses from the Holy Bible

Book	Chapter	Verse	Page Number
Matthew	13	24	63
Matthew	13	31-32	64
Exodus	35	10	64
Psalms	26	2	64

Appendix III

The human electromagnetic centres and their qualities

	ELECTROMAGNETIC CENTRES	QUALITIES
1	Coccyx (root)	Creative force Self preservation Seat of Kundalini
2	Solar plexus	Agent for desires, fears, ego, emotions Energised by solar magnetic waves Personal power
3,4,5	Heart (3 Nodes)	Luminous centre for love, healing and life distribution Divine intelligence
6	Thymus	Cellular healing response, repair and survival
7	Throat	Centre of communication and creative expression
8	Brain stem	Mind/body transmission Light activated
9	Third venticle (Limbic system)	Telepathy Clairvoyance/Clairaudient Intution, Insight Light activated
11,12	Crown (anterior fontanel grey layer)	Radar/antennae Connection with collective consciousness Complex thinking Light activated
13		**Source of Life**

Reference

Angha, Molana al-Moazam Hazrat Mir Ghotbeddin Mohammad. (1979). *Az Janin Ta Janan (From Foetus to Paradise)*. Tehran: M.T.O. Shahmaghsoudi® Publications.

Angha, Molana al-Moazam Hazrat Salaheddin Ali Nader (Professor Nader Angha). (2001). *Expansion & Contraction Within Being (Dahm)*. Riverside, CA: M.T.O. Shahmaghsoudi® Publications.

_____. (2011). *Sufism: Lecture Series*. London: MTO Publications®.

Angha, Molana al-Moazam Hazrat Shah Maghsoud Sadegh (Professor Sadegh Angha). (1987). *Al-Rasael*. Tehran: M.T.O. Shahmaghsoudi® Publications.

_____. (1998). *Chanteh – The Gnostic's Cosmos*. Riverside, CA: M.T.O. Shahmaghsoudi® Publications.

_____. (1982). *Hamaseh Hayat (The Epic of Life)*. Tehran: M.T.O. Shahmaghsoudi® Publications.

_____. (1938). *Eshghe-o-Sarnevesht (Love and Fate)*. Tehran: M.T.O. Shahmaghsoudi® Publications.

_____. (1997). *Manifestations of Thought*. Riverside, CA: M.T.O. Shahmaghsoudi® Publications.

_____. (1986). *Message from the Soul*. Verdugo City: M.T.O. Shahmaghsoudi® Publications

_____. (1986). *Mystery of Humanity,* Maryland: University Press of America® Inc.

_____. (1979) Owzan va Mizan (*Weights & Balances*). Tehran: M.T.O. Shahmaghsoudi® Publications

_____. (2005). *Weights & Balances*. Martinez, CA: M.T.O. Shahmaghsoudi® Publications

_____, Aryainejad S. (1994). *Professor Sadegh Angha's Theory of Particle Structure and Its Application*. New York: Vantage Press.

Badylak SF, Weiss DJ, Caplan A, Macchiarini P. (2012) Engineered whole organs and complex tissues. *The Lancet*. 379, 943-952.

Clune J, Pennock RT, Ofria C, Lenski RE. (2012) Ontogeny Tends to Recapitulate Phylogeny in Digital Organisms. T*he American Naturalist*. 180, E54-E63.

Davidson, A Herbert. (1992). *Alfarabi, Avicenna, & Averroes, on Intelect, Their Cosmologies, Theories of the Active Intellect, & Theories of Human Intellect*, New York: NY. Oxford University Press.

Eiseley, Loren. (1957). *The Immense Journey*. New York: NY. Random House Inc.

Eriksson PS, Perfilieva E, Björk-Eriksson T, Alborn AM,

Nordborg C, Peterson DA, Gage FH. (1998). Neurogenesis in the adult human hippocampus. *Nature Medicine.* 4, 1313-1317.

Flammarion, Camille. (1867). *Dieu Dans La Nature (God in Nature).* Paris: Didier et cie.

Gao J, Hooker BS, Anderson DB. Expression of Functional Human Coagulation Factor XIII A-domain in Plant Cell Suspensions and Whole Plants. *Protein Expression and Purification* 37, 89-96.

Hegel, Georg Wilhelm Friedrich, 1770-1831. (1977). *Hegel's Phenomenology of Spirit, Translated by A.V. Miller with Analysis of the Text and Foreword by J.N. Findlay.* New York, NY: Oxford University Press.

Kant, Immanuel, 1724-1804. (1987). *Critique of Judgment. Translated, with an Introduction by Pluhar W.S, with a Foreword by Gregor M.J.* Indianapolis, IN: Hackett Publishing Company, Inc.

Korobeinikova AV, Garber MB, Gongadze GM. (2012). Ribosomal proteins: structure, function, and evolution. *Biochemistry (Mosc).* 6, 562-574.

Leibniz, Gottfried Wilhelm, Freiherr von, 1646-1716. (1991). *Leibniz's Monadology: an edition for students / Nicholas Rescher.* Pittsburgh, Pa: University of Pittsburgh Press.

Materlinck, Maurice. (1899). *Wisdom and Destiny.* New York, NY: Dodd, Mead and Company.

Muller, Friedrich Max. (1890-92). *The Hymns of the Rig Veda with Sayana's Commentary, 2nd edition.* Oxford: Oxford University Press.

Spinoza, Benedictus de, 1632-1677. (2006). *The Essential Spinoza: Ethics and Related Writings / edited by Michael L . Morgan ; with translations of Samuel Shirley*. Indianapolis, IN: Hackett Publishing Company, Inc.

Strehler B, Hirsch G. (1971). Codon-restriction theory by aging and development. *Theoretical Biology*. 33, 429-474.

Vivekananda, Swami. (1982). *Jnana-Yoga*. New York, NY: Ramakrishna-Vivekananda Center.

Xenophanes, Lesher JH. (2001). *Xenophanes of Colophon: Fragments: A text and translation with a commentary by J.H. Lesher*. Toronto, Ontario: University of Toronto Press Incorporated.

Index

F

G

H

I

J

K

L

M

N

O

P

Q

R

S

T

Printed in Dunstable, United Kingdom

63869986R00127

BIOGRAPHY

Christi Caldwell is the *USA Today* best-selling author of the Sinful Brides series and the Heart of a Duke series. She blames novelist Judith McNaught for luring her into the world of historical romance. When Christi was at the University of Connecticut, she began writing her own tales of love—ones where even the most perfect heroes and heroines had imperfections. She learned to enjoy torturing her couples before they earned their well-deserved happily ever after. Christi lives in Charlotte, North Carolina where she spends her time writing, baking, and being a mommy to the most inspiring little boy and empathetic, spirited girls who, with their mischievous twin antics, offer an endless source of story ideas!

Visit www.christicaldwellauthor.com to learn more about what Christi is working on, or join her on Facebook at Christi Caldwell Author, and Twitter @ChristiCaldwell!

In Need of a Duke

A Prequel Novella to "The Heart of a Duke" Series

In Need of a Duke: (Author's Note: This is a prequel novella to "The Heart of a Duke" series by Christi Caldwell. It was originally available in "The Heart of a Duke" Collection and is now being published as an individual novella.

It features a new prologue and epilogue.

Years earlier, a gypsy woman passed to Lady Aldora Adamson and her friends a heart pendant that promised them each the heart of a duke.

Now, a young lady, with her family facing ruin and scandal, Lady Aldora doesn't have time for mythical stories about cheap baubles. She needs to save her sisters and brother by marrying a titled gentleman with wealth and power to his name. She sets her bespectacled sights upon the Marquess of St. James.

Turned out by his father after a tragic scandal, Lord Michael Knightly has grown into a powerful, but self-made man. With the whispers and stares that still follow him, he would rather be anywhere but London…

Until he meets Lady Aldora, a young woman who mistakes him for his brother, the Marquess of St. James. The connection between Aldora and Michael is immediate and as they come to know one another, Aldora's feelings for Michael war with her sisterly responsibilities. With her family's dire situation, a man of Michael's scandalous past will never do.

Ultimately, Aldora must choose between her responsibilities as a sister and her love for Michael.

For Love of the Duke

Book 1 in the "Heart of a Duke" Series

After the tragic death of his wife, Jasper, the 8th Duke of Bainbridge buried himself away in the dark cold walls of his home, Castle Blackwood. When he's coaxed out of his self-imposed exile to attend the amusements of the Frost Fair, his life is irrevocably changed by his fateful meeting with Lady Katherine Adamson.

With her tight brown ringlets and silly white-ruffled gowns, Lady Katherine Adamson has found her dance card empty for two Seasons. After her father's passing, Katherine learned the unreliability of men, and is determined to depend on no one, except herself. Until she meets Jasper…

In a desperate bid to avoid a match arranged by her family, Katherine makes the Duke of Bainbridge a shocking proposition—one that he accepts.

Only, as Katherine begins to love Jasper, she finds the arrangement agreed upon is not enough. And Jasper is left to decide if protecting his heart is more important than fighting for Katherine's love.

More Than a Duke

Book 2 in the "Heart of a Duke" Series

Polite Society doesn't take Lady Anne Adamson seriously. However, Anne isn't just another pretty young miss. When she discovers her father betrayed her mother's love and her family descended into poverty, Anne comes up with a plan to marry a respectable, powerful, and honorable gentleman—a man nothing like her philandering father.

Armed with the heart of a duke pendant, fabled to land the wearer a duke's heart, she decides to enlist the aid of the notorious Harry, 6th Earl of Stanhope. A scoundrel with a scandalous past, he is the last gentleman she'd ever wed…however, his reputation marks him the perfect man to school her in the art of seduction so she might ensnare the illustrious Duke of Crawford.

Harry, the Earl of Stanhope is a jaded, cynical rogue who lives for his own pleasures. Having been thrown over by the only woman he ever loved so she could wed a duke, he's not at all surprised when Lady Anne approaches him with her scheme to capture another duke's affection. He's come to appreciate that all women are in fact greedy, title-grasping, self-indulgent creatures. And with Anne's history of grating on his every last nerve, she is the last woman he'd ever agree to school in the art of seduction. Only his friendship with the lady's sister compels him to help.

What begins as a pretend courtship, born of lessons on seduction, becomes something more leaving Anne to decide if she can give her heart to a reckless rogue, and Harry must decide if he's willing to again trust in a lady's love.

The Love of a Rogue

Book 3 in the "Heart of a Duke" Series

Lady Imogen Moore hasn't had an easy time of it since she made her Come Out. With her betrothed, a powerful duke breaking it off to wed her sister, she's become the *tons* favorite piece of gossip. Never again wanting to experience the pain of a broken heart, she's resolved to make a match with a polite, respectable gentleman. The last thing she wants is another reckless rogue.

Lord Alex Edgerton has a problem. His brother, tired of Alex's carousing has charged him with chaperoning their remaining, unwed sister about *ton* events. Shopping? No, thank you. Attending the theatre? He'd rather be at Forbidden Pleasures with a scantily clad beauty upon his lap. The task of *chaperone* becomes even more of a bother when his sister drags along her dearest friend, Lady Imogen to social functions. The last thing he wants in his life is a young, innocent English miss.

Except, as Alex and Imogen are thrown together, passions flare and Alex comes to find he not only wants Imogen in his bed, but also in his heart. Yet now he must convince Imogen to risk all, on the heart of a rogue.

Loved By a Duke

Book 4 in the "Heart of a Duke" Series

For ten years, Lady Daisy Meadows has been in love with Auric, the Duke of Crawford. Ever since his gallant rescue years earlier, Daisy knew she was destined to be his Duchess. Unfortunately, Auric sees her as his best friend's sister and nothing more. But perhaps, if she can manage to find the fabled heart of a duke pendant, she will win over the heart of her duke.

Auric, the Duke of Crawford enjoys Daisy's company. The last thing he is interested in however, is pursuing a romance with a woman he's known since she was in leading strings. This season, Daisy is turning up in the oddest places and he cannot help but notice that she is no longer a girl. But Auric wouldn't do something as foolhardy as to fall in love with Daisy. He couldn't. Not with the guilt he carries over his past sins… Not when he has no right to her heart…But perhaps, just perhaps, she can forgive the past and trust that he'd forever cherish her heart—but will she let him?

To Love a Lord

Book 5 in the "Heart of a Duke" Series

All she wants is security:

The last place finishing school instructor Mrs. Jane Munroe belongs, is in polite Society. Vowing to never wed, she's been scuttled around from post to post. Now she finds herself in the Marquess of Waverly's household. She's never met a nobleman she liked, and when she meets the pompous, arrogant marquess, she remembers why. But soon, she discovers Gabriel is unlike any gentleman she's ever known.

All he wants is a companion for his sister:

What Gabriel finds himself with instead, is a fiery spirited, bespectacled woman who entices him at every corner and challenges his age-old vow to never trust his heart to a woman. But…there is something suspicious about his sister's companion. And he is determined to find out just what it is.

All they need is each other:

As Gabriel and Jane confront the truth of their feelings, the lies and secrets between them begin to unravel. And Jane is left to decide whether or not it is ever truly safe to love a lord.

THE HEART OF A SCOUNDREL

Book 6 in the "Heart of a Duke" Series

Ruthless, wicked, and dark, the Marquess of Rutland rouses terror in the breast of ladies and nobleman alike. All Edmund wants in life is power. After he was publically humiliated by his one love Lady Margaret, he vowed vengeance, using Margaret's niece, as his pawn. Except, he's thwarted by another, more enticing target—Miss Phoebe Barrett.

Miss Phoebe Barrett knows precisely the shame she's been born to. Because her father is a shocking letch she's learned to form her own opinions on a person's worth. After a chance meeting with the Marquess of Rutland, she is captivated by the mysterious man. He, too, is a victim of society's scorn, but the more encounters she has with Edmund, the more she knows there is powerful depth and emotion to the jaded marquess.

The lady wreaks havoc on Edmund's plans for revenge and he finds he wants Phoebe, at all costs. As she's drawn into the darkness of his world, Phoebe risks being destroyed by Edmund's ruthlessness. And Phoebe who desires love at all costs, has to determine if she can ever truly trust the heart of a scoundrel.

To Wed His Christmas Lady

Book 7 in the "Heart of a Duke" Series

She's longing to be loved:

Lady Cara Falcot has only served one purpose to her loathsome father—to increase his power through a marriage to the future Duke of Billingsley. As such, she's built protective walls about her heart, and presents an icy facade to the world around her. Journeying home from her finishing school for the Christmas holidays, Cara's carriage is stranded during a winter storm. She's forced to tarry at a ramshackle inn, where she immediately antagonizes another patron—William.

He's avoiding his duty in favor of one last adventure:

William Hargrove, the Marquess of Grafton has wanted only one thing in life—to avoid the future match his parents would have him make to a cold, duke's daughter. He's returning home from a blissful eight years of traveling the world to see to his responsibilities. But when a winter storm interrupts his trip and lands him at a falling-down inn, he's forced to share company with a commanding Lady Cara who initially reminds him exactly of the woman he so desperately wants to avoid.

A Christmas snowstorm ushers in the spirit of the season:

At the holiday time, these two people who despise each other due to first perceptions are offered renewed beginnings and fresh starts. As this gruff stranger breaks down the walls she's built about herself, Cara has to determine whether she can truly open her heart to trusting that any man is capable of good and that she herself is capable of love. And William has to set aside all previous thoughts he's carried of the polished ladies like Cara, to be the man to show her that love.

To Trust a Rogue

Book 8 in the "Heart of a Duke" Series

A rogue

Marcus, the Viscount Wessex has carefully crafted the image of rogue and charmer for Polite Society. Under that façade, however, dwells a man whose dreams were shattered almost eight years earlier by a young lady who captured his heart, pledged her love, and then left him, with nothing more than a curt note.

A widow

Eight years earlier, faced with no other choice, Mrs. Eleanor Collins, fled London and the only man she ever loved, Marcus, Viscount Wessex. She has now returned to serve as a companion for her elderly aunt with a daughter in tow. Even though they're next door neighbors, there is little reason for her to move in the same circles as Marcus, just in case, she vows to avoid him, for he reminds her of all she lost when she left.

Reunited

As their paths continue to cross, Marcus finds his desire for Eleanor just as strong, but he learned long ago she's not to be trusted. He will offer her a place in his bed, but not anything more. Only, Eleanor has no interest in this new, roguish man. The more time they spend together, the protective wall they've constructed to keep the other out, begin to break. With all the betrayals and secrets between them, Marcus has to open his heart again. And Eleanor must decide if it's ever safe to trust a rogue.

The Lure of a Rake

Book 9 in the "Heart of a Duke" Series

A Lady Dreaming of Love

Lady Genevieve Farendale has a scandalous past. Jilted at the altar years earlier and exiled by her family, she's now returned to London to prove she can be a proper lady. Even though she's not given up on the hope of marrying for love, she's wary of trusting again. Then she meets Cedric Falcot, the Marquess of St. Albans whose seductive ways set her heart aflutter. But with her sordid history, Genevieve knows a rake can also easily destroy her.

An Unlikely Pairing

What begins as a chance encounter between Cedric and Genevieve becomes something more. As they continue to meet, passions stir. But with Genevieve's hope for true love, she fears Cedric will be unable to give up his wayward lifestyle. After all, Cedric has spent years protecting his heart, and keeping everyone out. Slowly, she chips away at all the walls he's built, but when he falters, Genevieve can't offer him redemption. Now, it's up to Cedric to prove to Genevieve that the love of a man is far more powerful than the lure of a rake.

To Woo a Widow

Book 10 in the "Heart of a Duke" Series

They see a brokenhearted widow.

She's far from shattered.

Lady Philippa Winston is never marrying again. After her late husband's cruelty that she kept so well hidden, she has no desire to search for love.

Years ago, Miles Brookfield, the Marquess of Guilford, made a frivolous vow he never thought would come to fruition—he promised to marry his mother's goddaughter if he was unwed by the age of thirty. Now, to his dismay, he's faced with honoring that pledge. But when he encounters the beautiful and intriguing Lady Philippa, Miles knows his true path in life. It's up to him to break down every belief Philippa carries about gentlemen, proving that not only is love real, but that he is the man deserving of her sheltered heart.

Will Philippa let down her guard and allow Miles to woo a widow in desperate need of his love?

To Redeem a Rake

Book 11 in the "Heart of a Duke" Series

He's spent years scandalizing society.

Now, this rake must change his ways.

Society's most infamous scoundrel, Daniel Winterbourne, the Earl of Montfort, has been promised a small fortune if he can relinquish his wayward, carousing lifestyle. And behaving means he must also help find a respectable companion for his youngest sister—someone who will guide her and whom she can emulate. However, Daniel knows no such woman. But when he encounters a childhood friend, Daniel believes she may just be the answer to all of his problems.

Having been secretly humiliated by an unscrupulous blackguard years earlier, Miss Daphne Smith dreams of finding work at Ladies of Hope, an institution that provides an education for disabled women. With her sordid past and a disfigured leg, few opportunities arise for a woman such as she. Knowing Daniel's history, she wishes to avoid him, but working for his sister is exactly the stepping stone she needs.

Their attraction intensifies as Daniel and Daphne grow closer, preparing his sister for the London Season. But Daniel must resist his desire for a woman tarnished by scandal while Daphne is reminded of the boy she once knew. Can society's most notorious rake redeem his reputation and become the man Daphne deserves?

One Winter with a Baron

Book 12 in the "Heart of a Duke" Series

A clever spinster:

Content with her spinster lifestyle, Miss Sybil Cunning wants to prove that a future as an unmarried woman is the only life for her. As a bluestocking who values hard, empirical data, Sybil needs help with her research. Nolan Pratt, Baron Webb, one of society's most scandalous rakes, is the perfect gentleman to help her. After all, he inspires fear in proper mothers and desire within their daughters.

A notorious rake:

Society may be aware of Nolan Pratt, Baron's Webb's wicked ways, but what he has carefully hidden is his miserable handling of his family's finances. When Sybil presents him the opportunity to earn much-needed funds, he can't refuse.

A winter to remember:

However, what begins as a business arrangement becomes something more and with every meeting, Sybil slips inside his heart. Can this clever woman look beneath the veneer of a coldhearted rake to see the man Nolan truly is?

To Enchant a Wicked Duke

Book 13 in the "Heart of a Duke" Series

A Devil in Disguise

Years ago, when Nick Tallings, the recent Duke of Huntly, watched his family destroyed at the hands of a merciless nobleman, he vowed revenge. But his efforts had been futile, as his enemy, Lord Rutland is without weakness.

Until now…

With his rival finally happily married, Nick is able to set his ruthless scheme into motion. His plot hinges upon Lord Rutland's innocent, empty-headed sister-in-law, Justina Barrett. Nick will ruin her, marry her, and then leave her brokenhearted.

A Lady Dreaming of Love

From the moment Justina Barrett makes her Come Out, she is labeled a Diamond. Even with her ruthless father determined to sell her off to the highest bidder, Justina never gives up on her hope for a good, honorable gentleman who values her wit more than her looks.

A Not-So-Chance Meeting

Nick's ploy to ensnare Justina falls neatly into place in the streets of London. With each carefully orchestrated encounter, he slips further and further inside the lady's heart, never anticipating that Justina, with her quick wit and strength, will break down his own defenses. As Nick's plans begins to unravel, he's left to determine which is more important—Justina's love or his vow for vengeance. But can Justina ever forgive the duke who deceived her?

Beguiled by a Baron

Book 14 in the "Heart of a Duke" Series

A Lady with a Secret… Partially deaf, with a birthmark marring her face, Bridget Hamilton is content with her life, even if she's been cast out of her family. But her peaceful existence—expanding her mind with her study of rare books—is threatened with an ultimatum from her evil brother—steal a valuable book or give up her son. Bridget has no choice; her son is her world.

A Lord with a Purpose… Vail Basingstoke, Baron Chilton is known throughout London as the Bastard Baron. After battling at Waterloo, he establishes himself as the foremost dealer in rare books and builds a fortune, determined to never be like the self-serving duke who sired him. He devotes his life to growing his fortune to care for his illegitimate siblings, also fathered by the duke. The chance to sell a highly coveted book for a financial windfall is his only thought.

Two Paths Collide… When Bridget masquerades as the baron's newest housekeeper, he's hopelessly intrigued by her quick wit and her skill with antique tomes. Wary from having his heart broken in the past, it should be easy enough to keep Bridget at arm's length, yet desire for her dogs his steps. As they spend time in each other's company, understanding for life grows as does love, but when Bridget's integrity is called into question, Vail's world is shattered—as is his heart again. Now Bridget and Vail will have to overcome the horrendous secrets and lies between them to grasp a love—and life—together.

To Tempt a Scoundrel

Book 15 in the "Heart of a Duke" Series

Never trust a gentleman…

Once before, Lady Alice Winterbourne trusted her heart to an honorable, respectable man… only to be jilted in the scandal of the Season. Longing for an escape from all the whispers and humiliation, Alice eagerly accepts an invitation to her friend's house party. In the country, she hopes to find some peace from the embarrassment left in London… Unfortunately, she finds her former betrothed and his new bride in attendance.

Never love a lady…

Lord Rhys Brookfield has no interest in marriage. Ever. He's worked quite hard at building both his fortune and his reputation as a rogue—and intends to enjoy all that they can offer him. That is if his match-making mother will stop pairing him with prospective brides. When Rhys and Alice meet, sparks flare. But with every new encounter, their first impressions of one another are challenged and an unlikely friendship is forged.

Desperate, Rhys proposes a pretend courtship, one meant to spite Alice's former betrothed and prevent any matchmaking attempts toward Rhys. What neither expects is that a pretense can become so much more. Or that a burning passion can heal… and hurt.

OTHER BOOKS IN THE HEART OF A DUKE SERIES BY CHRISTI CALDWELL

TO HOLD A LADY'S SECRET

Book 16 in the "Heart of a Duke" Series

Lady Gillian Farendale is in trouble. Her titled father has dragged her through one London Season after another, until the sheer monotony of the marriage mart and the last vestige of Gillian's once-independent spirit conspire to lead her into a single night of folly. When her adventure goes so very wrong, she has only one old friend to whom she can turn for help.

Colin Lockhart's youthful friendship with Lady Gillian cost him everything, and a duke's by-blow had little enough to start with. He's survived years on London's roughest streets to become a highly successful Bow Street Runner, and his dream of his own inquiry agency is almost within his grasp.

Then Gillian begs him to once again risk angering her powerful father. The ruthless logic of the street tells Colin that he dare not help Gillian, while his tender heart tempts him to once again risk everything for the only woman he'll ever love.

Coming November 2023

Once Upon a Betrothal, brings readers a winter-set romance between former lovers, Mr. Broden Burgess and Miss Harmony Beresford.

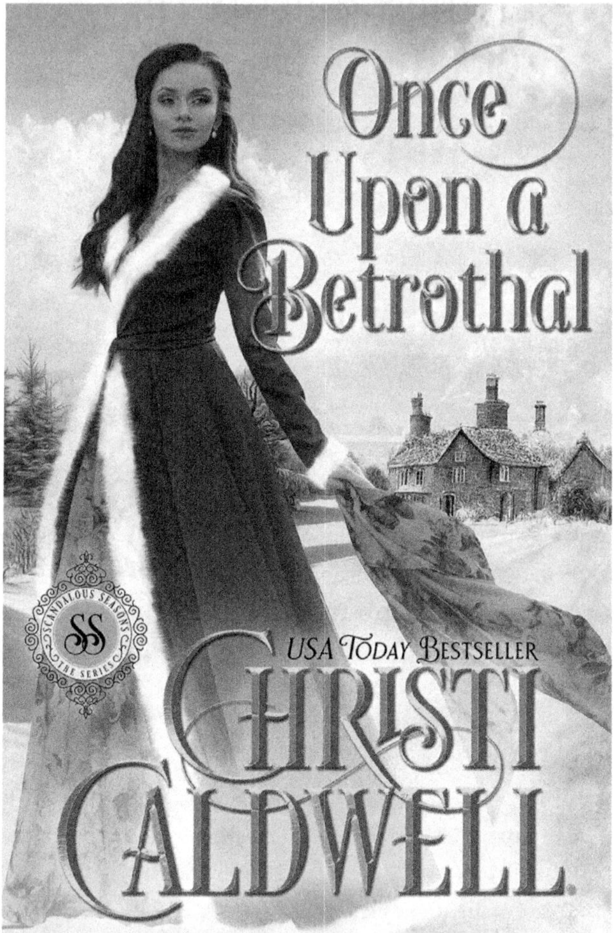

Also, join Christi Caldwell as she returns to where it all started—the Scandalous Seasons series!

Be swept away in the passion, drama, and splendor of Regency England, where the ballrooms glitter and the dukes are sexy and brooding, and the ladies dare to love.

Enjoyed *Endlessly Courted, Finally Loved*? Be sure and check out the next books coming in the Scandalous Seasons series!

Coming September 2023

Once a Rake, Suddenly a Suitor sees London's most scandalous rogue, Phineas Lesar, the Marquess of Landon falling for the last woman he should...Miss Anwen Kearsley, his best friend's younger sister.

because of this. So that fate could bring her together with this man, the most stubborn, wonderful, handsome Scot ever born.

THE END